Food-Energy-Water Systems Integration

This book analyses the challenges and opportunities within the food-energy-water nexus, offering actionable insights into workable resource management practices. By bridging the gap between practical utilisation and science, the book produces a link that integrates food, energy, and water to improve the sustainability of these natural resources. It offers a possible effective basis aimed at fostering innovative and collaborative research in this sector.

Features:

- Emphasises the sustainability of natural resources, encouraging readers to understand the importance of harmonising food, energy, and water demands for sustainable development.
- Provides methodologies and practical tools to measure and enhance nexus performance for resource optimisation across different locations.
- Integrates interdisciplinary perspectives, drawing from engineering, environmental science, economics, and policy research to provide a holistic view of the nexus.
- Examines the development of the nexus, offering insights into past trends, current conditions, and prospective states.
- Outlines actionable strategies for optimising the application and management of food, energy, and water resources.

The book is aimed at graduate and postgraduate students, researchers in environmental engineering, including policymakers, and industry practitioners seeking innovative approaches to resource management within the food-energy-water nexus.

Food-Energy-Water Systems Integration

Modelling, Optimisation and Sustainable Resource Management

Oliver O. Apeh and Nnamdi Nwulu

CRC Press
Taylor & Francis Group
Boca Raton London New York

CRC Press is an imprint of the
Taylor & Francis Group, an **informa** business

Designed cover image: Getty Images

First edition published 2025
by CRC Press
2385 NW Executive Center Drive, Suite 320, Boca Raton FL 33431

and by CRC Press
4 Park Square, Milton Park, Abingdon, Oxon, OX14 4RN

CRC Press is an imprint of Taylor & Francis Group, LLC

ISBN: 978-1-041-03758-3 (hbk)
ISBN: 978-1-041-06113-7 (pbk)
ISBN: 978-1-003-63384-6 (ebk)

DOI: 10.1201/9781003633846

Typeset in Times
by SPi Technologies India Pvt Ltd (Straive)

Dedication

To Our Families and Children

Contents

List of Acronyms

4IR	Integrating Fourth Industrial Revolution
ACLED	Armed Conflict Location and Event Data Project
APS	Announced Pledges Scenario
AR5	Fifth Assessment Report
BAFU	Bundesamt Für Umwelt
BCM	Billion Cubic Metres
CAP	Common Agricultural Policy
CBD	Convention on Biological Diversity
CCD	Climate-Compatible Development
COVID-19	Coronavirus Disease 2019
CRBFWP	Columbia River Basin Fish Wildlife Program
CRD	Climate Resilient Development
CSP	Concentrating Solar Power
DRC	Democratic Republic of the Congo
EDF	Electricité de France
EU	European Union
EV	Electric Vehicle
EWS	Early Warning Systems
FAO	Food and Agriculture Organization
FEW	Food-Energy-Water
FEWE	Food-Energy-Water-Ecosystem
GDP	Gross Domestic Product
GFWS	Global Food and Water System
GHG	Greenhouse Gases
GWL	Global Warming Level
IFPRI	International Food Policy Research Institute
INFEWS	Innovations at the Nexus of Food, Energy, and Water Systems
IPCC	Intergovernmental Panel on Climate Change
IPRA	International Peace Research Association
IRENA	International Renewable Energy Agency
IWRM	Integrated Water Resources Management
LCA	Life Cycle Assessment
MuSIASEM	Multi-Scale Integrated Analysis of Societal and Ecosystem Metabolism
NBS	Nature-Based Solutions
NEST	Nexus Solutions Tool
NGO	Non-Governmental Organisation
NSF	National Science Foundation
NZE	Net Zero Emissions
OECD	Organisation for Economic Cooperation and Development
PRIMA	Platform for Regional Integrated Modelling and Analysis
PV	Photovoltaics

REMap	Renewable Energy Map
SAPP	Southern African Power Pool
SDGs	Sustainable Development Goals
SLR	Systematic Literature Review
SSP3	Shared Socio-economic Pathway 3
STEPS	Stated Policy Scenarios
UK	United Kingdom
UN	United Nations
UNEP	United Nations Environment Programme
UNU	United Nations University
US	United States
USA	Urban and Ecosystem Facilities
WEF	Water-Energy-Food
WEFE	Water-Energy-Food-Ecosystems
ZAMCOM	Zambezi Watercourse Commission

Preface

The 21st century encounters many challenges, such as societal issues and constant environmental degradation, which result in significant threats to sustainable development. Prioritising health and family and promoting suitable environments appear to be dominant purposes. However, the occurrence of social distancing, the impending danger of pandemics, and rising environmental misuse aggravate these challenges. This causes frequent worsening of water systems, landscapes, ecosystems, and natural habitats—the basics of human life. The dominant sectoral concentration in academic research fails to contribute the broad knowledge needed to address these interrelated challenges efficiently. It is apparent that the quest for renewable energy loses its importance when water sources are polluted, and the potential of smart cities reduces as sea levels increase. Hence, recognising the nexus in which various elements are interconnected is essential.

In this book, the idea of the nexus, mainly the Food-Energy-Water (FEW) nexus, emphasises the interrelationship of crucial systems, enabling mutual support and effectiveness. Yet, further development remains a necessity. The present FEW nexus agenda falls short of stimulating confidence in a sustainable future where people cohabit harmoniously with nature. These sustainability challenges develop because of the complex interaction between environmental and societal activities at numerous levels. Therefore, modern methodologies are necessary for efficiently understanding and addressing these tangled social-ecological issues. The nexus strategy intends to irradiate the integrated nature of sustainability challenges, acknowledging synergies and trade-offs between several sectors of social-ecological systems. Although the nexus perspectives aim for possible scenarios, it remains uncertain which procedural strategies can successfully navigate the intricate dynamics of FEW nexus challenges and recognise interferences to address evolving phenomena.

Therefore, the book conceptualises the discussion between the nexus and climate change, stressing its socio-political scopes while addressing the importance of sustainability in the modern global ecological crises. Consequently, this relationship is an essential viewpoint in the search for Sustainable Development Goals (SDGs), given the acknowledgement of the necessity of interconnectivity between conventionally secluded sectors. Besides, nexus methods are unlimited to isolated regions; it is necessary to recognise and work from a local viewpoint in analysing open systems to the global community.

The book explores a broad research of case studies on contemporary issues surrounding Africa's food-energy-water-ecosystem (FEWE) nexus. This book studies the existing state through multidisciplinary perspectives, characterising the core challenges and possible solutions. It stresses the need for joint methods that consider the synergies and trade-offs among food, energy, water, and ecosystems to ensure resilience and unbiased access in the face of growing socio-economic and environmental challenges.

These focal themes highlight aligning the book's content with its main objectives. The book targets academics, policy experts, and researchers, including PhD and master's students across disciplines such as development planning, public policy, environment, local and urban studies, energy studies, public health, climate change, transportation, and environmental planning/engineering. This book intends to contribute to global discussions on leveraging agents for sustainable practices in food, energy, and water integration. We have curated this book to enrich dialogues on optimising and implementing sustainability measures. We anticipate that the breadth of knowledge presented within these studies will also serve as a valuable resource for instructors of undergraduate and graduate courses, making "FEW Nexus Optimisation and Climate Change" an engaging and informative addition to classroom curricula.

Authors' Biography

Oliver O. Apeh is a postdoctoral research fellow at the University of Johannesburg, South Africa. He has worked as a research assistant at the Fort Hare Institute of Technology (FHIT), University of Fort Hare, South Africa. He also has more than two decades of experience teaching science courses in high schools and higher institutions, as well as working as a laboratory demonstrator for university undergraduates.

He received BSc and MSc degrees in Physics and Astronomy from the University of Nigeria, Nsukka, in 2011 and 2015, respectively. Also, in 2014, he obtained a Postgraduate Diploma in Education (PGDE) from the National Open University of Nigeria (NOUN). He further obtained his PhD in Physics from the University of Fort Hare, Eastern Cape, South Africa, in 2021. His chief research interest covers aspects of renewable energy, including nanomaterial sciences, solar photovoltaic systems, smart energy systems, sustainable development studies, and data science for digital transformation. As an emerging researcher, he has published several peer-reviewed journal papers. His current research is on blockchain technology, food-energy-water nexus, and machine learning for solar energy.

Dr Apeh is a professional physicist registered with the South African Institute of Physics (SAIP). He is a member of the review teams for the following journals: *Journal of Digital Food Energy and Water Systems* (*JD-FEWS*), *Energy and Environment* (*EAE*), *IET Renewable Power Generation*, and *Recent Advances in Electrical & Electronic Engineering*. Moreover, he is one of the editors of the journals *American Journal of Modern Energy* and *Journal of Artificial Intelligence and Robotics*.

Nnamdi Nwulu is a Professor of Electrical Engineering and the Director of the Centre for Cyber Physical Food, Energy & Water Systems (CCP-FEWS) at the University of Johannesburg. His research interests have focused on applying digital technologies, mathematical optimisation techniques, and machine learning algorithms in food, energy, and water systems. This research focus has resulted in five books and over 70 indexed journal publications, earning him the 2021 NRF Research Excellence Award for Early Career/Emerging Researchers in the Engineering/Technology Category. He was also a 2020/2021 TW Kambule-NSTF-South32 Award finalist in the category of Emerging Researcher and won the 2020 UJ Vice Chancellor's Most Promising Young Researcher award. He is a member of the South African Young Academy of Science (SAYAS).

He is a professional engineer registered with the Engineering Council of South Africa (ECSA), a senior member of the Institute of Electrical and Electronics Engineers (SMIEEE), a senior member of the South African Institute of Electrical Engineers (SMSAIEE), and a Y-rated researcher by the National Research Foundation of South Africa.

He is the Editor-in-Chief of the *Journal of Digital Food Energy and Water Systems* (JD-FEWS) and Associate Editor of *IET Renewable Power Generation* (*IET-RPG*) and the *African Journal of Science, Technology, Innovation and Development* (*AJSTID*).

1 Background Information of the Book

1.1 BACKGROUND OF THE BOOK

Given that there are various crucial challenges the human race is presently encountering, it is pertinent to ask questions about the difficulties of choosing food, energy, and water for integration. Why not economics, land, and air? Why now, at the start of the 21st century? This book builds up by providing these essential questions. From a layman's perspective, food may be considered the most crucial component of the food-energy-water nexus. People can survive longer without food and energy than without water. Nevertheless, humans rely on the sun's energy as the primary energy source that guarantees humans' existence on the Earth's surface. Moreover, developed countries rely more on energy for a luxurious life and flexibility. This highlights the vital need of energy for humanity. The reality is that human beings require all these resources and cordial interactions which are described in this book.

Chemistry interconnects the three resources. Each of them is abruptly discoursed separately concerning their comparative status in society (imagine that the sun keeps shining on the Earth's surface indefinitely), and their connection with the other natural resources is described. The research has depicted that water, after air, is the most crucial need in society for human survival. In the same way, food is required for human survival over an extended period. Energy is required for supplying clean water and producing food crops. Sequentially, food is a source of energy. For instance, ethanol produced from corn and water is commonly utilised for energy production. The food, energy, and water (FEW) nexus is described in Figure 1.1.

All three resources are interconnected, and their sustainability within the nexus is equally important. This book examines the role of food, energy, and water in sustaining human survival and existence and emphasises the optimisation of the FEW nexus for the 21st century. Research associated with this topic is briefly described. The book deliberates the rate at which the three essential resources relate with one another for the common advantages of humans and how these connections can be managed to help humanity. It investigates the areas and deliberates techniques to advance the application of these resources with the least harm to the surroundings.

The integration of FEW systems is essential as humankind and the environment need all the three constantly, and each is interdependent. Thus, this book provides the collaborative nature of food, energy, and water, which relate to human beings and society. Besides, the book examines the human needs of FEW interactions and how ecosystems play significant parts, including policy and law, economics, climate change, and infrastructure, ensuring their relationships.

Moreover, this book separately analyses mechanisms for integrating food, energy, water, and ecosystems. The book deals with relevant questions, such as spatial and

DOI: 10.1201/9781003633846-1

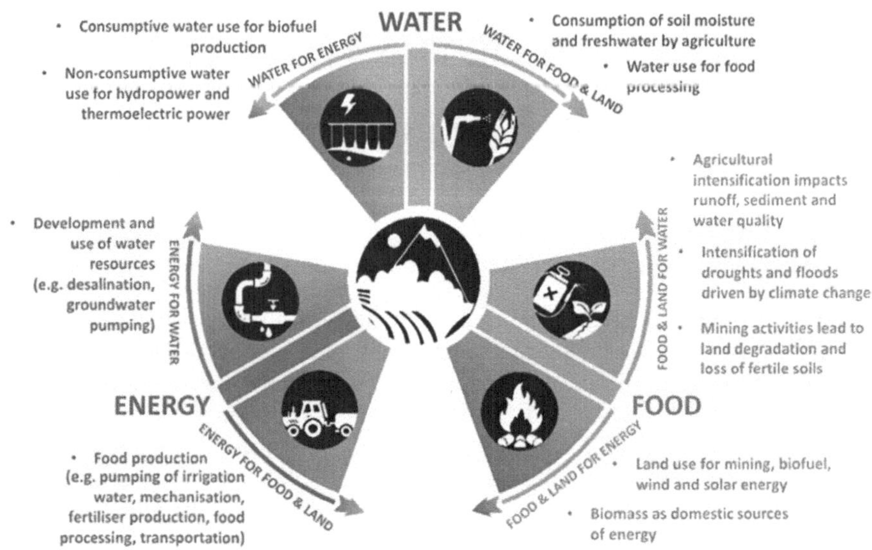

FIGURE 1.1 The relationship between food, energy, and water systems (United Nations, 2015).

time-based boundaries, that can produce results which could benefit policymakers in the real world as they concurrently reply to the problems confronting the demands for FEW, which are human and scientific challenges. While the FEW nexus is presently a developing area of scientific interest, it is not new. This book presents the state of the art where important studies performed over many decades ago have documented two fields: optimisation and sustainability.

1.2 FUNDAMENTAL CHALLENGES

While the FEW nexus comes with significant benefits in one aspect, it is crucial to remember that this system is not yet broadly recognised; so, it naturally comes with some challenges. According to research, the population and increased living standards have risen exponentially, necessitating increased demands for fundamental human needs such as food, energy, and water (David et al., 2022). Concerns about the availability of these three resources are serious, as they are essential for meeting the present and future demands. For example, recent studies estimate that approximately 768 million inhabitants worldwide lack access to safe water sources (WHO/UNICEF, 2022), and about 1.3 billion population remain without electricity supply (Arranz-Piera et al., 2018). Around 868 million people are underfed because of either food scarcity or the absence of food supply (Hamadani et al., 2020). Predictions for 2050 show that the demand for water worldwide (regarding water extractions) will rise by 55% (Boretti and Rosa, 2019), the demand for energy from hydropower and other renewable sources will increase by 60% (Popp et al., 2021), and the demand for supply of food will essentially rise by approximately 60% to meet the growing demands (Am et al., 2020). Climate change is expected to severely affect FEW systems around the world. Each of them will probably have intimidating costs for socio-economic growth and environmental sustainability. Considering these, there are pressing and urgent questions regarding the FEW systems, particularly the climate change scenario. As these questions have frequently and separately been discussed in the respective areas, such a regional method is insufficient because of the solid relations among the various regions. For example, (1) the share of global extractions of water for agricultural purposes is around 70%, (2) the application of energy production account for approximately 75% in industrial water extractions, (3) approximately 30% of world energy utilisation are used for food production and supply chain, and (4) around 90% of world power production is water-concentrated. Thus, a combined method is required to integrate the multiple connections among the various areas.

In this regard, the study of the "FEW nexus" has drawn significant interest in the present era, while the role of climate change is also increasingly integrated. Despite recent work and progress, knowledge of the FEW nexus remains restricted. Different parameters contribute to this: (1) the networks and the relationships are integrally complicated, and a robust scientific concept for researching such multipart energetically developing schemes and related networks is mainly unavailable, (2) there are natural and anthropogenic effects, many of which remain poorly understood, and (3) the data are frequently missing and incomplete. Also, other significant parameters have not received sufficient concentration. For example, understanding the fundamental meaning of the FEW nexus is needed. Studies on these are frequently dispersed and disjointed, making it difficult to discuss the FEW nexus intelligibly. This unusual subject is intended to gather some recent, but still dispersed, studies on the FEW nexus into a single forum to deliberate on the main challenges in researching this nexus and to explore opportunities for discussion (Apeh and Nwulu, 2024).

However, research studies on integration of FEW systems involve the use of different fields of study such as engineering; biological, physical, and social sciences; policy contribution, public health, and law; and knowledge of appropriate aspects of arts and literature. The authors do not aim to categorise the comparative significance of various aspects of contribution. Each has substantial value, and omitting any aspect can produce ineffective results in any specific condition. In the same way, the FEW nexus is generated by the collaboration of several social, biological, and physical systems, and effective contributions of the FEW nexus result from the combination of various relevant areas of academic effort. Throughout this book, users are encouraged to remember several challenging academic areas in this aspect. Food, energy, and water integrations are challenging areas in particular. The combination of all three into a network remains a developing area since it is particularly scientifically challenging, demanding cautious deliberation as listed below:

1. The central questions that drive all future scientific deliberations and the value of results.
2. Appropriate spatial and historical indicators are needed to assess data requirements, scientific and geographic considerations, and the needs of policymakers and other end-users.
3. The selections and costs involve selecting a measurement system that provides value to product customers, based on available data, models, and computational analyses.

4. How to select and integrate diverse datasets of varying quality.
5. How to select the correct category of the system and develop the precise system, based on all appropriate criteria that can combine the whole parts of a combined model, identify *all important* inner and outside connections, and generate meaningful outcomes for consumers with existing computational resources.

1.3 WHY FOOD, ENERGY, AND WATER?

Developing countries are expected to experience population growth and increased consumption, leading to significant resource stress. By 2050, there will likely be twice the demand for energy worldwide, whereas the demand for food and water is predicted to rise by more than 50% (IRENA, 2015). Furthermore, the impacts of climate change, including the rise in global temperatures and more frequent extreme weather conditions, increase the difficulty of addressing this growing demand. As the Earth approaches its sustainable resource limits, rivalries for these scarce commodities are expected to become even more critical.

Economic growth is expected to face increasing limitations due to the scarcity of one or more of these vital resources. Consequently, declaring food, energy, and water security has become a prominent challenge for national and international programmes for quite a while. Incorporating a "nexus" agenda that integrates food, energy, and water is inferred as a convincing approach towards advancing the Sustainable Development Goals (SDGs) and increasing resource efficiency. This step thoroughly interprets the results' effects within an area and the other parts involved. A key example illustrating the complex relationship within this nexus is the evolving trend of using biofuels as an energy source in transportation. The production of biofuels faces challenges, as it raises concerns regarding the allocation of limited water and land resources between biofuel generation and the cultivation of food crops for human consumption. An example of a complex relationship within the nexus is the growing adoption of biofuels as a primary energy source in the transportation industry.

Meanwhile, research on nexus highlights the importance of integrating FEW systems around five closely interconnected principles. These three resources are essential, highly productive, need massive infrastructure. They currently have significant impact, should be available and cheap, and could serve as the foundation for advanced policymaking and planning. The FEW nexus systems are studied and managed instead of individual FEW systems because they usually and exceptionally meet many or all of these five fundamental principles.

1.3.1 The Commodities Are Highly Demanded for Human Consumption

Human society needs a constant supply of food and water for consumption and survival, as well as energy to sustain many forms of pleasant life. Food, energy, and water are indispensable human resources for the existence of life and continuous renewal. These resources are crucial in both quantity and quality, such as nutritious food, efficient energy sources, and clean freshwater. The expression "society is only three meals away from anarchy" (sometimes "nine meals" or "from revolution") highlights the essential requirement for food. People can live without food, possibly for 30–40 days, and without water for 2–3 days. Water and food supplies depend on energy support, along with many other aspects of life that are essential for human existence. This is constantly observed during a power failure, when people experience complete disorder and dysfunction. Similarly, interruptions in water supplies have contributed to serious battle-affected areas in Iraq and Syria.

Regular incidences of power failures can demonstrate how communities can rapidly descend into dysfunction, approaching chaos, in the face of interruptions to modern energy supplies. Notably, in conflict-ridden countries such as Syria and Iraq, the criticality of water supply interruptions has been emphasised, while in 2018–2019, Venezuela faced devastating disruptions in energy supplies. The significance of these "three great consumables" is summarised in the extensively utilised words "food security," "energy security," and "water security," representing the indispensable means people and societies demand cheap and sufficient quantities of diet, energy, and water for strong and fruitful existence.

1.3.2 Availability and Consumption Trends of the Nexus Components

The increasing necessities for food, energy, and water are propelled by collective efforts, including international trade, climate change, population growth and mobility, technological and cultural changes, sustainable development, urbanisation, and changing lifestyles. However, if mismanagement of resources persists, it will undoubtedly result in a growing need. Nevertheless, the possibility of reducing this rising demand lies in decreasing wastage and losses in the food, energy, and water streams. This approach would also assist in recovering entrenched resources during production and reducing environmental effects. A decrease in energy and water use through conservation and effective usage is essential for the future generation.

1.3.2.1 Food

Concerns about the ability to meet the increasing food demand are rapidly growing worldwide. The availability of arable land and water for irrigation is diminishing, and these vital resources now face growing competition from industrialisation and urbanisation. The most generally accepted metric for assessing global food security is kcal/person/day. In 2005/2007, the global average per capita food availability for direct human consumption was 2,770 kcal/person/day (see Figure 1.3). However, this figure can be deceptive, as there are regions where this value falls below 2,500 kcal, while in others, it exceeds 3,000 kcal. The forecasts indicate that by 2050, world food production will need to increase by 60%, and developing countries will face an even more significant challenge with a required increase of 100% compared to 2005/2007 figures (Uhlenbrook and Connor, 2019). This corresponds to an annual growth rate of 1.1% in total global consumption. Figure 1.2 illustrates the projected values in million tons for crucial food groups relative to the 2005/2007 baseline. The driving forces behind this necessary increase will primarily stem from rising population numbers and income levels, as well as shifts in dietary patterns, such as a transition towards a more meat-based diet and the issue of overnutrition.

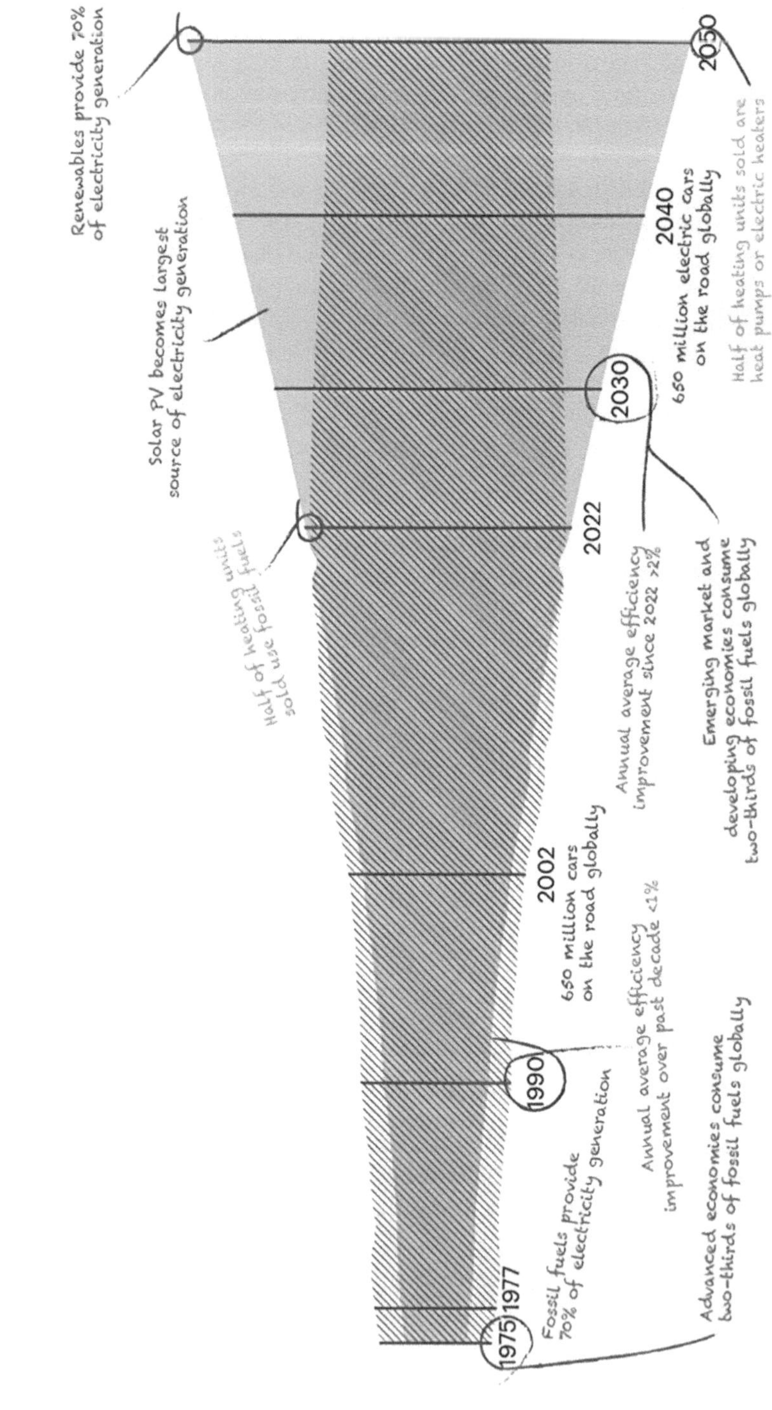

FIGURE 1.2 The projected values in million tons for crucial food groups relative to the 2005/2007 baseline.

1.3.2.2 Energy

The increase in energy demand is mainly propelled by features such as the rise in gross domestic product (GDP) and population growth. Notwithstanding different energy means, fossil fuels are expected to continue to be the primary fuel usage, establishing approximately 80% of energy production by 2035 (BP, 2016). Similarly, natural gas is likely to gain more acceptance in conjunction with the growing implementation of renewable energy, even though the contribution of the latter is projected to remain less than 10% by 2035. On the other hand, coal consumption is expected to decline, while oil use will remain stable. The extra demand for energy will originate from rising and developing economies, while the Organisation for Economic Cooperation and Development (OECD) countries are expected to demonstrate negligible growth.

Since 2010, there has been a decrease of more than 45% in the global population that lacks access to electricity. This development has been mainly motivated by improvements in developing Asia. However, nearly 760 million people live without electricity, as shown in Figure 1.3. The most critical condition exists in countries within sub-Saharan Africa (excluding South Africa), where roughly 80% of the population lacks electricity. While there have been some recent improvements in access in this region, they have not sustained it with population growth. Subsequently, there has been a 2.5% rise in populations lacking means since 2010. According to the stated policy scenarios (STEPS), the existing policies are expected to decrease globally, and the number of people without electricity access will be approximately 125 million by 2030. However, because of population growth, about 650 million

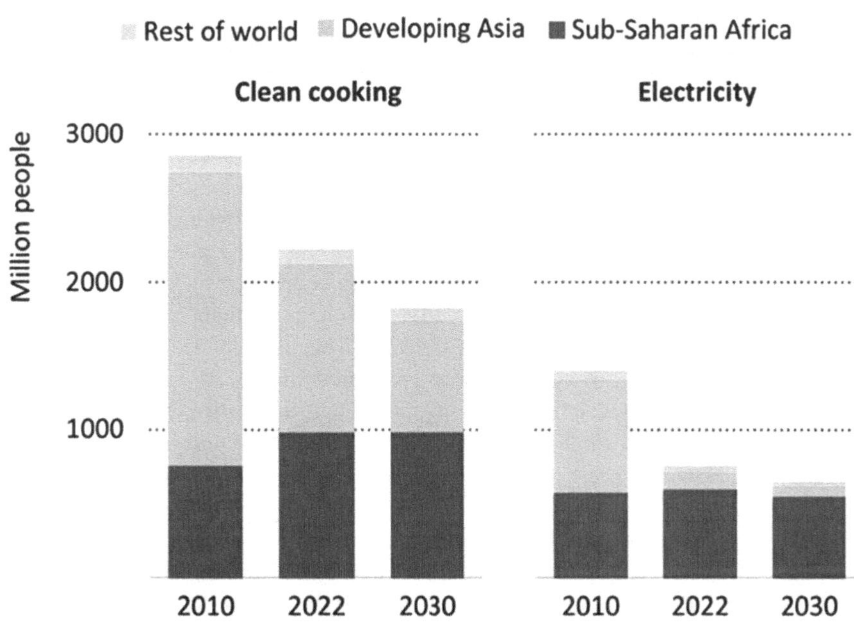

FIGURE 1.3 Population without access to modern energy in the STEPS.

individuals are expected to live without electricity by 2030 (Meyer et al., 2020; Apeh et al., 2022). In the announced pledges scenario (APS), the number of people without access to electricity is likely to reduce to 270 million by 2030. Moreover, in countries within sub-Saharan Africa, this number is anticipated to decrease by two-thirds. Lastly, in the net-zero emissions (NZE) scenario, common access to electricity is set to be attained by 2030.

Presently, almost 2.3 billion people rely on traditional kerosene, coal, or biomass for their cooking requirements, mainly in countries in developing Asia and across sub-Saharan Africa. Regarding the projection by the STEPS, the population living without access to clean cooking facilities is likely to be reduced by slightly over 15% by the year 2030. Suppose the outline goals of all countrywide clean cooking in the APS are met by 2030. In that case, the number of individuals lacking electricity will reduce by two-thirds, leaving roughly 735 million still without access. However, in the NZE scenario, universal access to clean cooking and electricity is predicted to be met by 2030.

In addition to meeting the increased energy demand necessary to support growing GDP in developing nations, there is a determined global effort to encourage electrification, hence safeguarding a stable growth in power generation. China will remain an essential player in forthcoming energy demand, as illustrated in Figure 1.3, demonstrating a shift towards a more sustainable route compared to earlier trends. As depicted in Figure 1.3, most energy estimates align with this trajectory and are supported by international initiatives focused on reducing reliance on fossil fuels. One prominent initiative is Sustainable Energy for All (SE4ALL), initiated in 2011 by the UN Secretary-General. The organisation has established three primary points agenda to be accomplished by 2030: ensuring common access to modern energy services, doubling the rate of energy efficiency increase in the global community, and doubling the contribution of renewable energy worldwide.

1.3.2.3 Water

Globally, a sufficient quantity of freshwater is available to meet world demands. However, some locations have been experiencing shortage of these reserves because of the uneven distribution, among other factors. According to the United Nations (UN) estimates, nearly 1.2 billion people live in regions undergoing physical water scarcity, while another 1.6 billion people experience economic water scarcity (Walsh et al., 2015). Regarding water quality, 748 million individuals cannot access an improved drinking water source (Pointet, 2022). This scarcity, in terms of quality and quantity, is expected to worsen due to growing demands, unsustainable extraction rates, deterioration of safe and quality water, and fluctuating weather patterns. Explicably, the primary influence of water scarcity is directly influenced by human consumption. However, there are indirect consequences on food production, ecosystems, and energy supply.

Usually, the increase in water resources was primarily driven by the needs of a growing population for sustenance, modern energy, and clothing. Recently, higher living standards among the middle-income class have led to rapid and significant spikes in water consumption for both production and personal use. Economic growth, coupled with better living conditions, may explain why the increase in water demand has outpaced population growth in the 20th century. In 2009, total global water withdrawals amounted to 4,500 billion cubic metres (BCM), with 70% allocated to agriculture, 17% to industrial use, and 13% to domestic and municipal uses (Young and Esau, 2015).

According to projections by the 2030 Water Resources Group (2009), water demand in 2030 under a business-as-usual scenario is expected to reach 6,900 BCM, which is 40% higher than the currently available, dependable, and sustainable water supplies (from both surface and groundwater sources). Another description by UN-Water suggests that water demand is expected to rise by approximately 55% in 2050, mostly propelled by sharp demands in thermal power plants, domestic usage, and the manufacturing sector (U. N. Water, 2018). The difference between demand and future availability cannot be bridged by finding additional water sources alone but will require operative demand-side management, necessitating robust policy interventions.

1.4 THE MEANING OF THE FEW NEXUS

The term nexus can be viewed to mean "to connect." The expression relates to the networks between two or more words, as they depend on one another or mutually. Thus, the FEW nexus is described as the relationships existing among the three resource elements alongside the trade-offs, interactions, and conflicts that might emerge from their managements, that is, food for water, water for food, food for energy, energy for water, and water for energy.

Different studies have reasoned that there are slight interactions on the FEW nexus with specific connotations, arguing with several opposing (and frequently overlapping) ideas (Benson et al., 2015; Al-Saidi and Elagib, 2017). Other researchers argue that the word is a buzzword, meaning an expression that gained importance because of "a mixture of an equivocal connotation and stout prescriptive resonance" (Cairns and Krzywoszynska, 2016). Further study suggests that several developing countries are unaware of the FEW nexus (Gain et al., 2015). Moreover, Cairns and Krzywoszynska (2016) observed that during the deliberations of the natural resource in the United Kingdom, the concept and utilisation of the word FEW nexus is "plural, disjointed, and unclear." Their worry is that the extensive usage of the word could underestimate its significance.

Wichelns (2017) suggests that the choice of food, energy, and water as the main elements of a nexus basis for research control and strategy, even though initially attractive, is pretty illogical (Wichelns, 2017). Studies present that while engineers and hydrologists speak of the water-energy-food (WEF) nexus, those in energy-related areas termed it the energy-water-food (EWF) nexus, and the agricultural domain uses the word FEW nexus. In terms of these differences in terminology, it becomes apparent that the ideological method relating to the FEW nexus primarily relies on the perception of specific policymakers or researchers (Bazilian et al., 2011). Allouche et al. (2015) affirm that the word can represent several elements to various people, suggesting that some researchers regard the FEW nexus sectors as shallow except for the environment and climate change, while other studies assess it as being reasonably comprehensive with a relationship to global resource security, poverty alleviation, and green economy (Pandey and Shrestha, 2017).

Water security has been the major challenge of the FEW nexus, as stated by the World Economic Forum, which different people call the WEF security nexus. Several alliances incorporating the FEW nexus methods have conflicting emphases, such as livelihoods, scarcity, sustainability, optimisation, the green economy, modelling, trade-offs, or climate. In the first four years of the Bonn 2011 Conference, Pahl-Wostl (2019) describes the robust concentration of FEW nexus on resource security. Still, later, the applications of the idea have extended to discourse mutuality and incorporation to attain the sustainable organisation of resources (Pahl-Wostl, 2019).

Even though there are differences in the meaning of "nexus," there have been many times when the scholar and developer have clashed in describing a word. Meadows and Randers (2012) note that 16 years after the Brundtland Commission mainstreamed the concept of sustainability, the international community had yet to agree on the meaning of the word. The discussion concerning the nexus' exact meaning and usage shows that it remains a developing idea (Allouche et al., 2015; Pandey and Shrestha, 2017). Even though there are contrary understandings of this context, de Loë and Patterson (2017) propose that "nexus thinking" remains dominant, in contrast to the exact stringent meaning of the FEW nexus (de Loë and Patterson, 2017).

1.5 REASONS BEHIND IMPLEMENTING FEW NEXUS AT THE BEGINNING OF THE 21ST CENTURY

Human activities have driven part of climate change, significantly affecting several environmental structures. These include changes in the global average temperature, as well as noticeable impacts, for example, the volume loss and evacuation of glaciers (Hugonnet et al., 2021), the development of new (peri-)glacial lakes (Mölg et al., 2021), seasonal changes, variations in snow lines, precipitation patterns, runoff, and evapotranspiration (Arnell, 2003; IPCC, 2021). To address these variations, Switzerland has introduced a series of progressive strategies at both national and cantonal levels, designed to alleviate and acclimate to climate change. The country first passed an Energy Act in 1998, subject to amendments afterwards, and accompanied by the outline of the "Swiss Energy Strategy 2050" in 2017. This broad policy tries to wipe away nuclear energy while eagerly encouraging the advancement of renewable energy sources, mainly hydropower. Moreover, they have actively followed policies for climate change, with national and cantonal devices being enforced in 2010 and 2012, respectively. These approaches are meant to coordinate the consumption of lakes, water reserves, subsurface, and surface waters (Zürich, 2018). Also, a national biodiversity policy was officially implemented in 2012 (BAFU, 2012), indicating Switzerland's commitment to addressing the broader environmental effects of climate change.

Concurrent with expanding innovative public strategies, other essential arrangements were developed to discourse climate change control instead of adaptation. The declining Trift glacier and the following establishment of a new lake encouraged the power company to plan the building of a new hydropower dam located 10 km away from Lake Grimsel. To address the continuous Grimsel dispute and knowing that non-governmental organisations (NGOs) might obstruct the plan, they introduced a participatory process to articulate an 80-year franchise, including stakeholders who could likely raise petitions (e.g., NGOs and affected local communities). The main emphasis of this enterprise was to equate biodiversity conservation and energy generation.

Unlike the region surrounding Lake Grimsel, the project zone is not presently under a defending position, and environmental NGOs would have a limited legal backup to obstruct the development. In this situation, driven by the commitment to increasing renewable energy for controlling climate change, environmental NGOs exhibited a strong readiness for collaboration. They determined that the formation of hydropower in the Trift region could be environmentally protected (Kellner, 2019). Nevertheless, plans leaning towards acclimatising, which try to organise water consumption across different areas, were not considered. The preliminary concession was submitted to the canton in 2017. Hitherto, as of early 2023, it had not been approved because of regulations. Remarkably, the hydropower company omitted a compulsory project execution timeline from the draft concession, and the canton of Bern did not include the project in its Cantonal Structure Proposal. However, the concession is expected to be approved soon. This concession can preserve and supply water without limitations, save for the lowest volume of residual water at the instantaneous discharge of the reservoir. This leads to a prospect for flexible hydropower generation, allowing the supply of energy during winter months and reinforcing the constancy of the electricity grid without considering water shortage downstream.

The procedures controlling the lake consider residual water and flood protection, but not water shortage conditions affecting downstream usages. When the lakes lack adequate inflow from upriver sources, it can cause short-term low overflow and elevated water temperatures downstream in the River Aare in dry seasons, parallel to historical occurrences (Kaderli, 2021). These occurrences could interrupt the stability of ecosystems, such as a decline in water-dependent biodiversity, benefiting generalist and warm water species while possibly presenting new non-native species, eventually causing less diverse societies with reduced functionality (FOEN, 2021). However, to protect water-bound biodiversity by guaranteeing suitable overflow, the cantonal authorities could ban or limit water extraction for irrigation. Farmers in Seeland are developing strategies for new irrigation projects to address rising temperatures and fluctuating precipitation patterns. They aim to employ more water from the River Aare to irrigate their farmland.

Studies have discovered that they are unaware of possible future restrictions on extraction because of lower Aare runoff, which could cause a loss of harvests and income for farmers during dry years. Accordingly, the unawareness has not resulted in disputes concerning upstream water procedures, especially about the planned 80-year concession for the hydropower dam. Despite being unified within a water cascade that depends on a similar upstream water supply and uses, the FEW systems function without coordination, following definite sectoral methods. This discrepancy is apparent in Figure 1.4, where ecological action situations are connected through the cascade, while social action situations are disconnected. Social-ecological action scenarios, which supervise water utilisation, do not allow different usages.

Figure 1.4 demonstrates a weak combination of water and energy sectors and unrepresentative harmonisation between food production and water management.

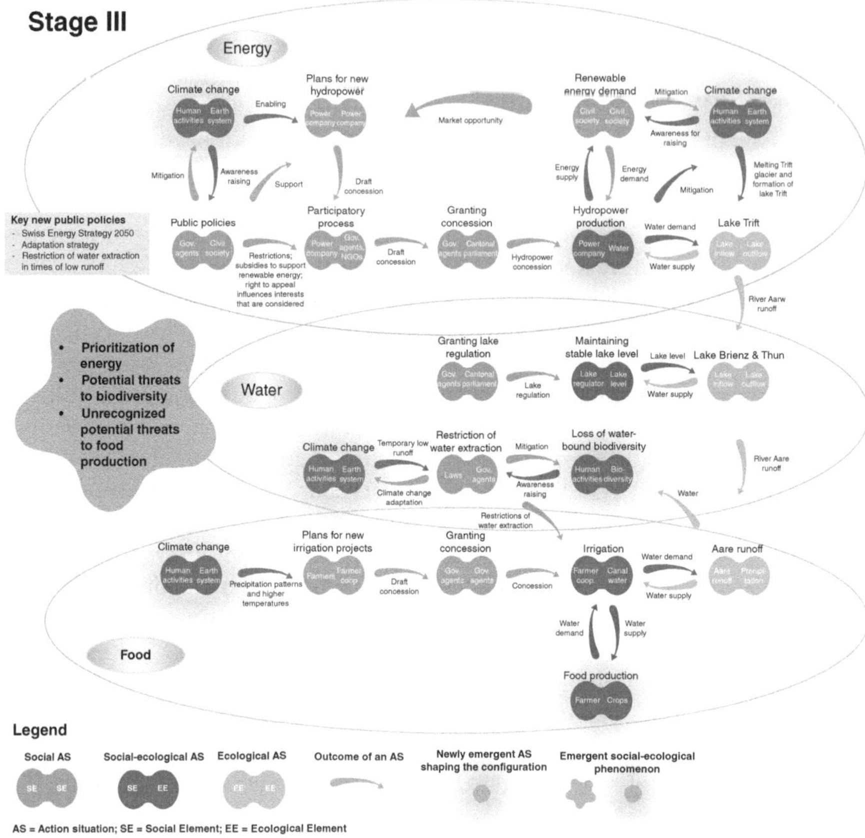

FIGURE 1.4 Illustration of the FEW nexus in the early 21st century through the Trift dam constructions as a case study. (Adapted from Kellner and Martin, 2023.)

The growing challenge of climate change demanded approaches for both regulation and acclimatisation, escalating the demand for hydropower and thus elevating the need to favour energy generation. This change in significance can risk biodiversity reliant on water ecosystems and pose unexpected dangers to food production. The growth of the FEW nexus phenomenon places increased emphasis on energy production, possibly endangering water-dependent biodiversity and creating unforeseen pressures on food production.

1.6 SEGMENTATION OF FOOD, ENERGY, WATER, AND CLIMATE SECURITY

Pachauri et al. (2009) proposed a novel research work termed the sectorisation of security, which discourses policy challenges surrounding food, energy, water, livelihood, climate security, and health. The authors emphasised that these definite security sectors can be investigated from several perspectives and with various focal points.

For example, water security holds several meanings for providers, local governments, and multinational water companies, primarily concerned with maximising prices while supplying low water quality standards. On the other hand, consumers advocate for a continuous supply of clean and potable water at a reasonable cost.

Similarly, health security encounters similar issues, mainly when the suppliers are pharmaceutical enterprises and multinational health organisations (World Health Organization, 2007). Health security is directly connected to water security, as many infant mortalities in the Global South stem from contaminated water sources and result in diarrhoeal ailments. Consumers need dependable and available healthcare services that are unlimited in their financial resources or burden them with permanent liability. The idea of food security, primarily articulated by Mechlem in 2004, was designed to eradicate malnutrition and starvation (Mechlem, 2004). However, global food institutions prefer to sell sub-standard food at inflated costs and employ regulation over the supply of essential grains to the Global South by applying genetically modified organisms. Their current method of food security intensely depends on production strategies, chemical inputs, and technology; hence, small-scale manufacturers in the South and North have become reliant on these technologies. Regularly, these crops fail in terms of the nutritional quality production needed for healthy living.

Moreover, this industrialised method of agriculture has several adverse impacts, including greenhouse gas (GHG) emissions associated with long-distance food transportation, soil depletion, and the weakening of small-scale farmers in the Global South due to subsidised agriculture in the Global North. These features jointly initiate a change in the food security pattern towards green agricultural food sovereignty, as supported by Via Campesina (2016) and strengthened by FAO (2016).

In a similar manner, energy security shares the same structural challenges. Multinational enterprises and oil manufacturers are frequently concerned about placing oil prices, irrespective of environmental effects like global warming and GHG discharges. On the other hand, consumers regularly demand a clean and dependable energy supply, preferably sourced from renewables, while also considering affordability. The idea of security cuts across five dimensions, including food, energy, water, ecosystem and health. Also included in these dimensions are political, economic, and military security, of which every aspect significantly affects environmental security because of pollution. This interconnection likewise spills over into communal security and can immediately endanger the safety of concerned populations, as presented in Figure 1.5.

To bolster these sectoral securities, an international change towards biofuels for energy security can affect food security, either by increasing the cost of food imported into countries or through changes in land usage from food crops to biofuel generation. Fuel production usually comprises deeply industrialised and mechanised agriculture, influencing soil preservation and water security. Moreover, excessive agrochemical usage in this process can contaminate soil, aquifers, and groundwater, with dangerous health impacts for livestock and humans. Also, biofuels have contributed extensively to food price increases, causing peaceful unrest and unstable social protests. This describes that food security is not only a global security challenge but also a human and national one. Widespread droughts across Argentina, Australia, and

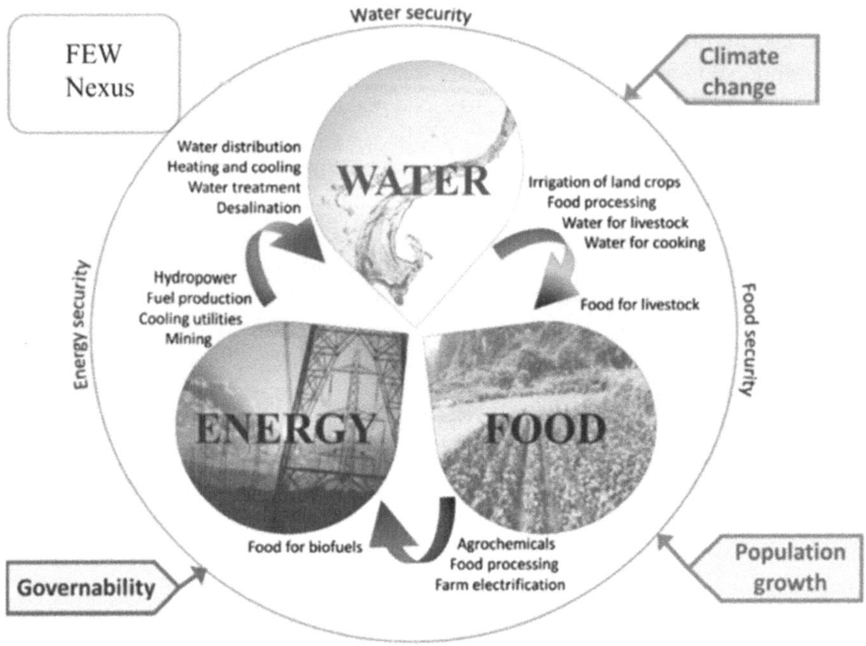

FIGURE 1.5 Outline of the food-energy-water (FEW) nexus security outfits. (Adapted from Mahlknecht et al., 2020.)

the United States in 2003 almost tripled the costs of basic grains, pressuring the FAO to ask the United States to restrict corn usage for biofuel production to ensure food supply and normalise grain costs in importing countries. Water security also causes regional problems, such as conflicts between downstream and upstream nations, regional droughts, accidents, and pollution. In dangerous situations, it can even result in military interference to protect the water supply in downstream countries. Nevertheless, a hydro-diplomatic method (Spring and Rodríguez, 2005; Gleick and Heberger, 2014) has been chiefly utilised in addressing transboundary water disputes (Wolf et al., 2003).

Health security is analysed from an international viewpoint, especially when primary outbreaks such as Asian flu, SARS, Asian flu, and AHN1 endanger the world's population because of fast transmission aided by modern transportation. In addition, viruses possibly used as weapons to perform terror become a national security danger to different countries, including the United States and Europe. Health security is mainly a human security challenge that affects individuals and families, particularly women who regularly assist as main caregivers for the sick.

- **Widening**: Five dimensions: societal, political, environmental (expanded security), military (narrow), and economic.
- **Deepening** (from people to state and global: levels and actors).
- **Sectorisation** (Food, energy, water, health, biodiversity security, and soil).

TABLE 1.1
Spreading, Extending, and Sectorisation of Security (Oswald Spring, 2020)

Security dimension ⇒ Level of interaction ⇓	Military	Political	Economic Financial	Environmental	Societal
Human/Gender individual & humankind ⇒			Food sec. Health sec.	Cause and victim	Food sec. Health sec. Water sec.
Societal/Community	DRR			⇩ ⇧	
National	Shrinking		Energy and water security	⇩ ⇧	Food and health security
International Regional			Water and food security	⇩ ⇧	Water security
Global/Planetary			Health security	Global environmental/ climate change	Health security

Each reference point and dimension point needs a detailed examination, which will be gradually addressed in subsequent chapters. Table 1.1 illustrates how these five security dimensions interconnect and can be utilised at various stages.

It is essential to highlight that the security method typically centres on one's global attitude and observations, inspired by leading media descriptions, governments, and dominant scientific knowledge. Notably, scholars have opposing stances, with critical theorists supporting a broader, more profound, and more sector-centred idea of security. They criticise the modern security researchers, mainly in the United States, who prefer the independent state as the essential referent body and assess the military as the leading securitising agent, accounting for protecting national territory and sovereignty.

1.7 THE SUSTAINABILITY CHALLENGE

The International Peace Research Association (IPRA) comprises researchers from various fields of study devoted to systematically analysing sustainable development plans locally (Oswald Spring and Oswald Spring, 2020). Also, they research conflict resolution and investigate broad, local, and regional practices that help this kind of progress. This research attempt seems to have paramount meaning, especially in a place where development and the interests connected with global agendas often have authority. Given this pressure the world is experiencing, any form of alternative development must prioritise human knowledge and its complicated interaction with the environment in this developmental search. These scholars promote policies that encourage social, regional, and economic growth and can restore inequality and suppress violence. At the same time, they pursue to enhance the application of natural resources, protecting them for generations to come while avoiding waste generation and pollution. This effort centres on implementing clean production processes, encouraging reasonable wealth circulation, eliminating poverty, sanitising the stewardship of natural resources, promoting family planning, and fostering civic co-responsibility, all focused on human rights values (Oswald Spring and Oswald Spring, 2020), solidarity, and respect. The technique encourages the organisation of urban and regional spaces, each within the basis of collective societal involvement and co-responsibility (Oswald Spring and Oswald Spring, 2019). In the 21st century, sustainable development reorganises the relationship between humanity, nature, and society, ensuring well-being and fairness (Velenturf and Purnell, 2021). This commitment is made in a framework controlled by multinational impact and globalisation, focused on global solidarity and collective global welfare in protection and existence.

The FEW resources are essential for human survival, and their production, purification, circulation, and consumption techniques are nearly intertwined (Yuan and Lo, 2022). The complex connections among these resources are usually called the FEW nexus. For example, food and energy production, especially in areas lacking irrigation, require insufficient water resources. At the same time, energy is critical for tasks including treating wastewater, pumping water for irrigation, and managing, processing, and conveying food items. Many significant challenges, such as climate change, urbanisation, and population increase, are interrelated and jointly pose an escalating risk to the practicality and constancy of FEW systems. Figure 1.6 illustrates a theoretical background showing the relationship between climate change and population growth within the FEW nexus.

The depicted model in Figure 1.6 explains the interaction between the FEW nexus and critical stressors, including population growth and climate change. In turn, these stressors exert mutual effects on FEW systems. The fundamental challenge lies in developing balanced strategies (such as mitigation or adaptation) that tolerate the sustainable application of global, regional, and local resources while meeting societal and environmental requirements. These challenges are expected to become especially severe in more urbanised and thickly populated countries by 2050. For instance, the accessibility of FEW resources in megacities centres on supply chains. Interruptions in the local supply of FEW resources threaten the sustainability of these cities. Furthermore, they cause issues of justice and social equity, connected by the unequal distribution of costs and reimbursements related to FEW resources.

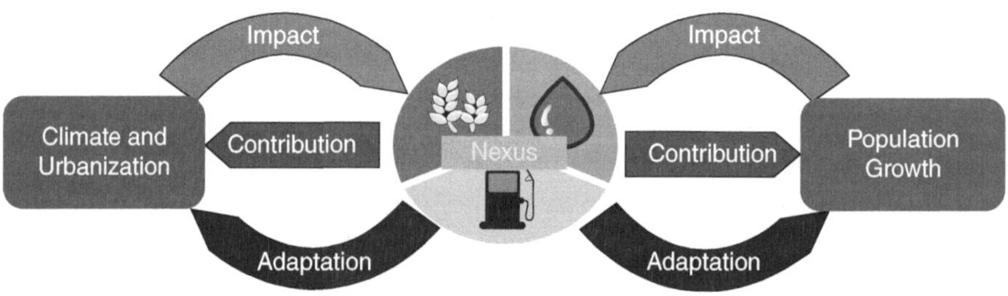

FIGURE 1.6 The interaction of climatic change and population growth within the FEW nexus domains.

Also, inequalities develop between private decisions and public regulatory rules concerning FEW resources, possibly marginalising vast parts of the populace.

Notably, when observing from input–output perspectives, there are sociological and ecological effects of growing FEW systems. These range from socio-economic and demographic changes to impacts on landfill operations, waste management, and air and water quality. Simplifying or forecasting how climate migration and demographic change influence FEW systems is hard. Eventually, the valuable ecosystem services rely on maintaining an equilibrium between the production and consumption of FEW resources. Nevertheless, the biodiversity loss, discharge of pollutants, severe interruptions in nutrient cycling, and rise in land use gradually challenge the functionality and flexibility of FEW systems vital for human sustenance. The interpretation of research discoveries from the areas of science and technology into solutions for FEW difficulties presents a considerable challenge. Differences in climates affect energy utilisation, water availability, and food supply chains while concurrently changing the capacity of ecosystems to meet human FEW needs and societal activities intended to strengthen ecosystem services. The specific way these dynamics rearrange models of business determinations, public initiatives, supply chains, and regulatory oversight remains imperfectly understood and mostly lags behind developments in science and technology. Suppose international innovation and cooperative solutions are lacking. In that case, severe stages may reach where ecosystem losses and environmental degradation become irreversible, indicating terrible costs for the rapid expansion of the global population. However, the knowledge and insights gathered through the China–U.S. Joint Research Center for Ecosystem and Environmental Change (JRCEEC) furnish a nice model pattern for collaborative research into the difficulties and consequences of shifting FEW systems, alongside possible remedies to challenges that enhance sustainable progress.

1.8 FEW NEXUS APPROACHES FOR SUSTAINABLE DEVELOPMENT

The nexus approach describes an integration method between governance and management across sectors and scales. The UN's SDGs (such as zero hunger, clean water and sanitation, and no poverty) are directly connected to the FEW nexus. Therefore, energy-related decisions that do not consider the trade-offs these decisions might have on food and water resources could have adverse effects. The nexus approach provides a valuable, transformative adaptation decision support tool for integrated resources management. Nexus planning also offers the potential to monitor progress towards achieving SDG targets. "It is all about resource security for sustainable development." Data scarcity at different spatial scales could be a significant limitation to the success of nexus planning. There is also a need to upscale nexus planning as a decision-making tool. This can help leverage the implementation process and progress towards SDGs and subsequent monitoring and evaluation simultaneously. The nexus approach essentially allows integration during the planning stage, which could mitigate some of those trade-offs. So, it is beneficial for investment planning to use that infrastructure to catalyse economic development and job creation.

1.9 CONCLUSION

Essential ideas for instituting and developing the difficulties in international transdisciplinary collaborations have emerged in the process of FEW nexus developments. Firstly, it is vital to respect individuals, their unique research methods, and the various national research processes where it is conducted. This respect enhances trust, allowing collaborative research for mutual development. Secondly, it is essential to address the challenges of every collaborator considerately and ensure a strong understanding. This is because research challenges and prize structures vary across countries. For instance, a hypothesis-driven method for a critical study is not necessarily the same as a mission-adapted problem-solving strategy. Besides, language barriers regularly confuse deliberations surrounding crucial research aims, clarifying experimental methods, drawing meaningful deductions, assessing result effects, and planning pathways for distributing findings. A severe third aspect comprises understanding the current research endeavours and strategic proposals of international partners. Mutual understanding of teaching attitudes, research abilities, interests, intellectual property sharing, and cultural differences assists in team formation, inspiring discourse, analysing essential research questions, and creating avenues for protecting institutional support for networking initiatives.

REFERENCES

Allouche, J., Middleton, C. and Gyawali, D. (2015) 'Technical veil, hidden politics: Interrogating the power linkages behind the nexus'. *Water Alternatives*, 8(1) pp. 610–626.

Al-Saidi, M. and Elagib, N. A. (2017) 'Towards understanding the integrative approach of the water, energy and food nexus'. *Science of the Total Environment*, 574, pp. 1131–1139.

Am, J. B. et al. (2020) *Innovation in a crisis: Why it is more critical than ever*. McKinsey & Company.

Apeh, O. O., Meyer, E. L. and Overen, O. K. (2022) 'Contributions of solar photovoltaic systems to environmental and socioeconomic aspects of national development—A review'. *Energies*, 15(16), p. 5963.

Apeh, O. O., & Nwulu, N. I. (2024). The water-energy-food-ecosystem nexus scenario in Africa: Perspective and policy implementations. *Energy Reports*, 11, 5947–5962.

Arnell, N. W. (2003) 'Effects of IPCC SRES* emissions scenarios on river runoff: A global perspective'. *Hydrology and Earth System Sciences*, 7(5), pp. 619–641.

Arranz-Piera, P. et al. (2018) 'Mini-grid electricity service based on local agricultural residues: Feasibility study in rural Ghana'. *Energy*, 153, 443–454.

BAFU (2012) 'Anpassung an den Klimawandel in der Schweiz. Ziele, Herausforderungen Und Handlungsfelder. Erster Teil der Strategie des Bundesrates vom 2. März 2012'. Bundesamt für Umwelt (BAFU) Bern.

Bazilian, M. et al. (2011) 'Considering the energy, water and food nexus: Towards an integrated modelling approach'. *Energy Policy*, 39(12), pp. 7896–7906.

Benson, D., Gain, A. K. and Rouillard, J. J. (2015) 'Water governance in a comparative perspective: From IWRM to a nexus approach?' *Water Alternatives*, 8(1), pp. 756–773.

Boretti, A. and Rosa, L. (2019) 'Reassessing the projections of the world water development report'. *NPJ Clean Water*, 2(1), p. 15.

BP. (2016). *BP Energy Outlook 2016: Outlook to 2035 (Spencer Dale & Bob Dudley, Foreword)*. London: BP p.l.c.

Cairns, R. and Krzywoszynska, A. (2016) 'Anatomy of a buzzword: The emergence of "the water-energy-food nexus" in UK natural resource debates'. *Environmental Science & Policy*, 64, pp. 164–170.

Campesina, V. (2016) 'La Via Campesina, building an international movement for food and seed sovereignty'. https://viacampesina.org/en/2016/02/la-via-campesina-building-an-international-movement-for-food-and-seed-sovereignty

David, L. O. et al. (2022) 'Integrating fourth industrial revolution (4IR) technologies into the water, energy & food nexus for sustainable security: A bibliometric analysis. *Journal of Cleaner Production*, 363, 132522. Doi: 10.1016/j.jclepro.2022.132522

De Loë, R. C. and Patterson, J. J. (2017) 'Rethinking water governance: Moving beyond Water-centric perspectives in a connected and changing world'. *Natural Resources Journal*, 57(1), pp. 75–100.

FAO. (2016) 'Agriculture organization, 2014'. In *Livestock primary*. Food and Agriculture Organization of the United Nations.

FOEN (2021) 'Effects of climate change on Swiss water bodies. Hydrology, water ecology and management'. In Federal Office for the Environment (FOEN) (Ed.), *Environmental Studies (No. 2101)* (p. 125). Federal Office for the Environment FOEN.

Gain, A. K., Giupponi, C. and Benson, D. (2015) 'The water–energy–food (WEF) security nexus: The policy perspective of Bangladesh'. *Water International*, 40(5–6), pp. 895–910.

Gleick, P. H. and Heberger, M. (2014) 'Water conflict chronology'. In Peter H. Gleick (Ed.), *The world's water: The biennial report on freshwater resources* (pp. 173–219). Springer.

Hamadani, J. D. et al. (2020) 'Immediate impact of stay-at-home orders to control COVID-19 transmission on socioeconomic conditions, food insecurity, mental health, and intimate partner violence in Bangladeshi women and their families: An interrupted time series'. *The Lancet Global Health*, 8(11), pp. e1380–e1389.

Hugonnet, R. et al. (2021) 'Accelerated global glacier mass loss in the early twenty-first century'. *Nature*, 592(7856), pp. 726–731.

Intergovernmental Panel on Climate Change (IPCC). (2021). Climate Change 2021: Impacts, adaptation, and vulnerability. *Contribution of Working Group II to the Sixth Assessment Report of the Intergovernmental Panel on Climate Change*. Cambridge University Press

IRENA (2015) *2030: Roadmap for a renewable energy future*. International Renewable Energy Agency, Abu Dhabi.

Kaderli, R. (2021) 'Reglemente für die Zukunft'. In *Untersuchung der Seeregulierung de s Brienzer-und Thunersees unter den Klimaszenarien CH2018 und Suche nach eine m optimalen Reglement für den Thunersee*. ETH Zurich.

Kellner, E. (2019) 'Social acceptance of a multi-purpose reservoir in a recently deglaciated landscape in the Swiss Alps'. *Sustainability*, 11(14), p. 3819.

Kellner, E. and Martin, D. A. (2023) 'Learning from past coevolutionary processes to envision sustainable futures: Extending an action situations approach to the water-energy-food nexus'. *Earth System Governance*, 15, p. 100168.

Mahlknecht, J., González-Bravo, R. and Loge, F. J. (2020) 'Water-energy-food security: A Nexus perspective of the current situation in Latin America and the Caribbean'. *Energy*, 194, p. 116824.

Meadows, D. and Randers, J. (2012) *The limits to growth: The 30-year update*. Routledge.

Mechlem, K. (2004) 'Food security and the right to food in the discourse of the United Nations'. *European Law Journal*, 10(5), pp. 631–648.

Meyer, E. L., Apeh, O. O. and Overen, O. K. (2020) 'Electrical and meteorological data acquisition system of a commercial and domestic microgrid for monitoring PV parameters'. *Applied Sciences (Switzerland)*, 10(24), pp. 1–18. doi: 10.3390/app10249092

Mölg, N. et al. (2021) 'Inventory and evolution of glacial lakes since the little ice age: Lessons from the case of Switzerland'. *Earth Surface Processes and Landforms*, 46(13), pp. 2551–2564.

Oswald Spring, Ú. (2020) *Earth at Risk in the 21st century: Rethinking peace, environment, gender, human, water, health, food, energy security, and mitigation, humanities, science, engineering*. http://www.springer.com/series/15230

Oswald Spring, Ú. and Oswald Spring, Ú. (2019) 'Peace and sustainability in a globalised World'. *Úrsula Oswald Spring: Pioneer on gender, peace, development, environment, food and water: With a foreword by Birgit Dechmann* (pp. 201–223). Springer.

Oswald Spring, Ú. and Oswald Spring, Ú. (2020) 'Peace and sustainability in a globalised world'. In *Earth at risk in the 21st century: Rethinking peace, environment, gender, and human, water, health, food, energy security, and migration: With a foreword by Lourdes Arizpe Schlosser and a preface by Hans Günter Brauch* (pp. 43–57). Springer.

Pachauri, R. K. et al. (2009) *Facing global environmental change: Environmental, human, energy, food, health and water security concepts*. Springer Science & Business Media.

Pahl-Wostl, C. (2019) 'Governance of the water-energy-food security nexus: A multi-level coordination challenge'. *Environmental Science & Policy*, 92, pp. 356–367.

Pandey, V. P. and Shrestha, S. (2017) 'Evolution of the nexus as a policy and development discourse'. *Water-energy-food nexus: Principles and practices* (pp. 11–20). Wiley Online Library.

Pointet, T. (2022) 'The United Nations world water development report 2022 on groundwater, a synthesis'. *LHB*, 108(1), p. 2090867.

Popp, J. et al. (2021) 'Bioeconomy: Biomass and biomass-based energy supply and demand'. *New Biotechnology*, 60, pp. 76–84.

Spring, Ú. O. and Rodríguez, M. de L. H. (2005) *El valor del agua: Una visión socioeconómica de un conflicto ambiental*. El Colegio de Tlaxcala.

U. N. Water (2018) *2018 UN world water development report, nature-based solutions for water*. UNESCO.

Uhlenbrook, S. and Connor, R. (2019) *The United Nations world water development report 2019: Leaving no one behind*. UNESCO.

United Nations (2015) *Reconciling resource uses in transboundary basins: Assessment of the water-food-energy- ecosystems nexus in the Sava River Basin* (pp. 1–121). UNECE. http://www.unece.org/fileadmin/DAM/env/water/publications/WAT_Nexus/ece_mp.wat_46_eng.pdf

Velenturf, A. P. M. and Purnell, P. (2021) 'Principles for a sustainable circular economy'. *Sustainable Production and Consumption*, 27, pp. 1437–1457.

Walsh, B. P., Murray, S. N. and O'Sullivan, D. T. J. (2015) 'The water-energy nexus, an ISO50001 water case study and the need for a water value system'. *Water Resources and Industry*, 10, pp. 15–28.

WHO/UNICEF (2022) *Progress on drinking water, sanitation and hygiene in schools*. https://data.unicef.org/wp-content/uploads/2022/08/JMP-WASH-in-schools_2022.pdf

Wichelns, D. (2017) 'The water-energy-food nexus: Is the increasing attention warranted, from either a research or policy perspective?' *Environmental Science & Policy*, 69, pp. 113–123.

Wolf, A. T., Yoffe, S. B. and Giordano, M. (2003) 'International waters: Identifying basins at Risk'. *Water Policy*, 5(1), pp. 29–60.

World Health Organization (2007) *The world health report 2007: A safer future: Global public health security in the 21st century*. World Health Organization.

Young, M. D. and Esau, C. (2015) 'Charting our water future: Economic frameworks to inform decision-making 1'. In Michael D. Young and Christine Esau (Eds.), *Investing in water for a green economy* (pp. 45–57). Routledge.

Yuan, M.-H. and Lo, S.-L. (2022) 'Principles of food-energy-water nexus governance'. *Renewable and Sustainable Energy Reviews*, 155, p. 111937.

Zürich, B. K. (2018) 'Klimawandel im Kanton Zürich'. In Building Department of the Canton of Zurich (Ed.), *Massnahmenplan Anpassung an den Klimawandel. Zürich*. Building Department of the Canton of Zurich.

2 The Evolution of Food-Energy-Water Nexus into a Sustainable System

2.1 INTRODUCTION

The world has experienced deep and consistent transitions in human and environmental systems since the mid-20th century, bringing about significant changes in global activities (Pant, Adhikari and Bhattarai, 2015) and the shift into a modern geological age—the Anthropocene (Lewis and Maslin, 2018). The planetary population tripled between 1950 and 2022, rising from 2.5 to 7.6 billion, with development focusing primarily on urban areas, quintupling from 751 million to 4.2 billion urban inhabitants (Chen et al., 2022). Together with these fantastic demographic changes, the world economy grew from US$ 9.25 to US$108 (Lutz, Butz and Samir, 2017). Alongside, the utilisation of food, energy, and water supplies to the universe rose steeply (Schlör, Venghaus and Hake, 2018), causing a reformation of the earth's matter–energy equilibrium (Williams et al., 2016) and living biomass (Elhacham et al., 2020). Nevertheless, substantial growth and progress in education, health, wealth, science, technology, and quick growth signify an experiential danger to human existence and global well-being (Masson-Delmotte et al., 2021). Human activities have changed the earth's environment and brought about wide damage in biodiversity, desertification, resource reduction, alteration of land use, as well as water, soil, and air pollution (Convention on Biological Diversity, 2020), forcing humanity outside different terrestrial borders (Lade et al., 2020). Simultaneously, there are uneven developmental activities where 770 million persons have no means to access electricity (IEA, 2021), 2.1 billion to secure and promptly accessible water (WHO/UNICEF, 2022), starvation remains the principal source of sickness universally (Fanzo et al., 2018), and an estimate of five billion people will be affected by the scarcity of water by 2050 (Water, 2018).

In contrast to this context, the idea of the FEW nexus developed in 2009–2011 has kept advancing and become effective and realised. Food, energy, and water resources are regarded as allied "systems" and "sectors," with the rising needs for them regularly seen from the perspective of "security" (Bizikova et al., 2013). Now, natural resource organisation and human growth meet as subjects of health and welfare, fairness and equity, and sustainability are comprehended as the social basis of the terrestrial top limit (Friedman et al., 2018) and a "secure and fair passage for publics and the planet." Therefore, the nexus has been progressively applied to evaluate

DOI: 10.1201/9781003633846-2

possible universal energisers of variation, interactions, and balances across segments (De Laurentiis, Hunt and Rogers, 2016) and expand authority through clear arrangement and administration. As the nexus establishes itself as a reliable framework in research, policy, and strategic planning, while continuing to evolve, there is an increasing demand to review its developmental, theoretical, and practical applications in connection to sustainable growth, Studies such as those of Glass and Newig (2019) have emphasised the implications of governance, particularly in organisation, incorporation, strategic unity, and political and social science viewpoints (Divides, 2010), for attaining sustainable development.

2.2 SUSTAINABLE, EQUITABLE, AND DIVERSE DEVELOPMENT APPROACHES

The difficulty relating to the natural resources in the global community requires deep thoughts and new solutions, which can connect a wide-ranging idea, including sustainable development, to worldwide standards, including fairness, peace, democracy, and justice. As a strategy to achieve security for all, the world is gradually unified and, it is formed by the growth of market services, always depicted as the only way to scientific advancement, greater productivity, and improved security, productivity, security, and human well-being. In just two decades of slowed growth in the developing countries, the number of deaths has risen tremendously because of water shortages, starvation, and disasters in the environment, unemployment, insecurity, and violence. This has led to condemnation of the neoliberal pattern in both advanced sectors and traditional and Christian settings.

The practices of bureaucratic socialism have devastated confidence in humanity's intrinsic goodness, revealing the severe authenticity of dictatorial power sustained exclusively by its presence. The failure of the incompetent communist world, together with the negative propensities of market capitalism, which regularly disregards concepts of fairness, sustainability, and overall well-being, emphasises the shortcomings of past doctrines. A new method is needed. As the last century drew nearer, the "end of history" concept shows the destruction of social commitment, indicating the success of transnational capitalism determined exclusively by profit, uninterested in human welfare and nature. This reality is starkly evident in the plight of immigrants from religious and fratricidal conflicts, as well as populations displaced by environmental degradation, such as desertification, salinisation, and environmental pollution.

The increasing effects of sea-level rise due to the greenhouse effect, along with illnesses caused by exposure to toxic substances and ozone layer depletion, all result from the uncontrolled exploitation of natural resources in search of profit maximisation. These issues emphasise the urgent demand for global strategies that confront these challenges and are free from conceptual unfairness or class justifications. To prevent the adverse effects highlighted, practical devices must be implemented with necessary short-term measures while offering development plans aligned with environmental and social imperatives on local, regional, and global scales. When contemplating the future of the planet, including regions such as Latin America or Mexico, it is crucial to critically examine past experiences, freeing them from inappropriate expectations and hidden interests. This process lays the foundation for a new environmental and socio-economic framework, one that is broad, intricate, pluralistic, automated, and peace-oriented.

Indeed, the global community has projected a diverse direction in current decades, emulating foreign hegemonic doctrines prioritising individual economic freedom over considerations of socio-environmental justice. Although the world and regional plans may prioritise equal opportunities, they often fail to ensure equality in political, social, economic, and living conditions. The fundamental challenge lies in a past disagreement shaped by a select group of transnational elites, who enforce a unilateral future on society, reducing individuals to the sheer subject of manipulation, lacking purpose or historical context. The strategies of the 1990s, emphasising market liberalisation and competition, gradually departed from the course of democracy, harmonisation, contribution, and general welfare.

The crucial proof of the ongoing anti-democratic growth is evident in the growing marginalisation of excluded groups from policymaking, broadening the gap between the state and its citizens. This brings about a quiet extinction in their suffering and deficiency. The neoliberal doctrine intensifies this prohibition across six dimensions: social, economic, cultural, gender, political, and environmental. At the same time, direct misuse continues, where social and human costs are ignored. This comprises the systemic insight against womenfolk, replicated in meagre earnings and insufficient political representation, as well as environmental degradation. The ozone layer depletion, declining freshwater resources, overfished seas, increasing ocean dead zones, desertification, biodiversity loss, climate change, and unchecked population growth have collectively produced a global inequality on Earth. Considering these challenges, utopia assumes an intricate function, possibly uniting and enriching rather than separating. The contributions generated by 180 countries in Rio de Janeiro in 1992, which took effect in 1997, reflect the acknowledgement of the complex interaction between anthropogenic and natural forces. Scientists and policymakers must collaborate to discover a sustainable solution for the prospective planet, humanity, and ecology.

Society, in general, must engage in this effort, reasoning globally but acting locally. The government cannot single-handedly address all the human and ecological challenges, ranging from resource depletion, waste accumulation, pervasive pollution, climate change, and depletion of the ozone layer; only through collaborative efforts of government and society, where both parties are committed to a brighter future, can these challenges be confronted. Consequently, with these efforts, peace, ecological security, enhanced social justice, quality of life, democracy, and gender equality could be achieved, especially in the third millennium (Millennium Development Goals, 2000). This transformative process must begin at home and extend to the stratosphere.

2.3 REGIONALITY GROWTH AND SUSTAINABILITY

Similar to Mexico, other biologically diverse nations in Latin America, as well as China and numerous Asian and African countries, possess huge prospects in harnessing their natural resources. Mexico is enriched with biodiversity because of the connection between the neotropical and neo-arctic ecosystems. Effective resource management could help alleviate current poverty and generate clean, dignified employment opportunities if ecological parameters are associated with economic planning. Natural resources must be preserved to boost social interactions, promoting better democracy and equal contribution across every aspect of humanity. Strengthening networks between ecology, economy, quality of life, and scientific-technological growth creates a new geopolitical pattern (Licha, 1996). This pattern comprises the unification of scientific knowledge, traditional wisdom, and morals to improve new perceptions and target to promote individuals prepared to confront the challenges of the upcoming era in ethical, economic, social (Ravetz, 2020), and personal scopes. This economisation can make production processes more resourceful, depending on immediate strategies promoting greater equality and equity. Thus, ecology evolves into a lucrative, limited asset, where natural resources formed over millions of years, acquire concrete value and are governed by market dynamics of supply and demand.

Efficient and reasonable natural resource management enables considerable energy savings, reduced pollution, and measurable economic benefits (Apeh, Meyer and Overen, 2022). On the other hand, alternative sources, including wind, water, and solar energy, coupled with improvements in energy conservation technologies, solid waste recycling, and efficient combustion systems, pose new challenges. These challenges determine a distinct pattern of globalisation grounded in security, equality, democracy, and sustainability. Besides, this method contributes to overall savings that extend beyond purely economic domains, and so moderates the risks of financial crises. Unfortunately, a significant risk arises in the fields of alternative biotechnology, energy efficiency, green technology, substitution of hazardous materials, and recycling, where most of the patents are held by multinational conglomerates (Jamison and Rohracher, 2001). This phenomenon worsens the inequality between the Global South and North, as global amounts and trade terms adversely lean towards developed countries and technology, services, and capital assets, while disadvantaging developing countries and their raw materials.

However, thermoelectric generation, burning of barrels of oil, atomic fusion, genetically modified organisms, and various methods of resource control, both past and present, have caused exact subtle environmental fluctuations worldwide, advancing both the present and future planet. For instance, if the global community had a comparable equal discharge each as Germany or the United States—and, since 2007, likewise China—had remained in 2018, we would require resources of about five Earths to overcome the adverse impacts of these discharges. Nevertheless, nations such as Mexico, characterised as threshold or middle-range nations endowed with vast biodiversity, qualified scientists creating valuable information, and native culture historically expert in preserving delicate ecosystems, may map out a course towards biodiversity protection (Dodds and Bartram, 2016). Protecting their indigenous culture centres on the complete and active contribution of all the people (Ron, 2011), motivated by ingenuity and democracy. Native and peasant sets of people, primarily women, possess the capacity to rehabilitate devastated ecosystems, thereby reducing the exploitation of natural resources that endanger current biodiversity and encourage industrialisation.

On a global measure, the global environmental conferences in Stockholm deliberated on sustainable development in 1972, in Rio de Janeiro on biodiversity in 1992, and in Kyoto finalised a contract form on GHGs in 1997. Also, the Basel Protocol on toxic waste, enforced in 1992, emphasises worldwide environmental challenges and the urgency for prompt resolutions. Eventually, the idea of human ecology (Handl, 2012) has created innovative concepts by connecting individual and social development patterns. This begins with a devoted entirety, where health, well-being, and global solidarity rest on fresh values. Simultaneously, it challenges Western nations' habits and cultural biases (ethnocentrism), where more significant income regularly comes at the expense of free time and relaxation. More geographic movement overlaps with destroying human connections, social obligations, family bonds, and community ties (Bryceson and Vuorela, 2020). Paid labour has succeeded the functions that were earlier contented within families for their sustenance. Subsistence agriculture (von Werlhof, 2008) and small-scale production produce opportunities for alternative employment and soil management sustainability, possibly enhancing food safety. Achievement and social mobility should not compromise the quality of individual relationships, which can bring about stress-related phenomena and reduced leisure time. Therefore, sustainable development encounters the challenges in interpersonal relations, economy, quality of life, and health to promote social mobility and happiness not at the cost of other countries or nature but through unity and careful resource management, marked by temperance. Applications from various global conferences are united, such as "small is beautiful"; "think globally, act locally"; "less consumerism, more quality of life"; "justice and peace for all".

This characteristic system of globalisation, separated from dominant and international processes, is expected by counterparts in the Southern Hemisphere, much like their counterparts in the Northern Hemisphere. To clarify more, Northern societies struggle with a tendency towards depersonalisation, stress-related ailments, immunodeficiency, time scarcity, and compromised quality of life. In contrast, the South contends with abject poverty, inhumane conditions, joblessness, low earnings, severe discrimination, and an increasing gap between social groups that brings about conflict and societal displeasure. Only through the basis of political, individual, economic, social, and environmental sustainability, promoting more gender equality across the range of family, community, nation, continent, and planet can achieve a genuine quality of life.

2.4 THE CHALLENGES TOWARDS SUSTAINABILITY

The IPRA brings together researchers from different disciplines, united in their commitment to systematically examine policies for sustainable development and conflict resolution within different local contexts (Oswald Spring and Oswald Spring, 2019). Their emphasis ranges from examining regional and local creativities that foster such development. In today's world, where unplanned policymaking and global plans are frequently controlled, prioritising human welfare and its interrelation with nature in any alternative development agenda becomes imperative. These researchers promote policies encouraging social, regional, and economic growth to alleviate violence and inequality, exploit natural resource assistance for future generations, and reduce waste and pollution through clean production systems. These efforts ensure equitable wealth distribution and poverty alleviation while efficiently managing natural resources.

Besides, they approve birth control measures and support civic co-responsibility entrenched in human rights values, solidarity, and respect (Oswald Spring and Oswald Spring, 2019). Promoting organised urban and territorial planning and inclusive societal participation is also stressed (Hendriks, 2006). As these researchers propose, sustainable development seeks to redefine the relationship between nature, humanity, and society, fostering prosperity and equity amidst globalisation and transnational influences in the 21st century. Central to this vision is global solidarity and a shared commitment to preservation and survival.

2.5 THE EVOLVING PICTURE AND TIMELINE OF THE NEXUS ACTIVITIES

Nexus has evolved since its inauguration (Lazaro et al., 2022). The stripling period of nexus is described as a "buzzword" and was used to define its policy (Cairns and Krzywoszynska, 2016). Nexus has undergone several modifications, bringing about various explanations, and the courses lack a general description (Endo et al., 2017). However, the nexus is usually defined as a bodily unified, interrelating, and complicated group of reserve systems containing food, energy, and water sectors with governance and socio-economic scopes. The reality is that the nexus is developing and, not surprisingly described, enhancing its flexibility. Segmental basics are frequently replaced or increased, producing a few fresh nexus frameworks, including "food-energy-land-water" (FELW) (Ringler, Bhaduri and Lawford, 2013), "water-land-energy-carbon" (WLEC) (Zhao et al., 2018), "food-energy-water-land use-climate" (FEWLC) (Laspidou, Mellios and Kofinas, 2019), "climate-land-energy-water systems" (CLEWs) (Ramos et al., 2021), or the "food-energy-water-ecosystem services" (FEWES) (Hejnowicz et al., 2022) nexus. Following these examples, a social-ecological method is beneficial to prevent "inflexibility traps" when highlighting relationships between model devices, such as land usage and climate change. More modifications comprise the food-water-energy-biodiversity (FWEB) nexus, which reinforces the relationship between food supply and biodiversity (Subedi, Karki, and Panday, 2020); the FEWE nexus, which describes the importance of solid ecosystems (Carmona-Moreno, Dondeynaz, and Biedler, 2019); and the food-energy-water-health (FEWH) nexus that pinpoints the connection between human health, resource usage, and social welfare in a global economic perspective (Slorach et al., 2020). The current development of the water-food-labour (WFL) nexus tries to distinguish the labour and social significance of consumption and production from an income and scarcity viewpoint (Distefano et al., 2022). The extensive application of the nexus may depend on its dynamic concept, tolerating several disciplinary sectors, areas, and policy fields to discover its vibrancy. Though nexus categories have evolved, a complete understanding of non-biophysical parts is still developing with additional foundation research. Generally, natural and physical science methods remain the major category of theoretical viewpoints (Albrecht, Crootof, and Scott, 2018). Consequently, these regularly disregard or undermine deliberation of political and social developments, restricting the evaluation of the nexus into an extensive quantifiable perspective.

These methods also fail to simplify a complete knowledge of how nexus networks work and are rooted in political frugality. In the context of growth overview, individual elements mostly viewed as non-political, such as issues related to food and land ownership, right to assets, and tenancy, are naturally not taken into account while investigating the nexus. The growing demand for a joint system of study and cross-disciplinary studies may assist in amending this inequality (Albrecht, Crootof, and Scott, 2018). The research interest indicates that while the basis is natural science-focused, the nexus study is starting to move outside the stipulated biophysical techniques to link with other categories. Nevertheless, many of the papers recognised were entirely from political or social science. This condition is contradictory because the nexus perception originated from political debate focused on the imperious discourse of food, energy, and water security matters, mainly outlined as significant economic and social growth problems. The security description arose as a strategy to retort to discourse about rising pressures on natural resources (WEFWI, 2011), including those in the Mekong Region (Lebel and Lebel, 2018). This description suggests that these pressures stimulate alleged "nexus insecurities," which means risks across the interconnected food, energy, and water systems. These risks are related to the reliability of various areas, their operational performance, and future development, all within a framework of bigger uncertainty (Dodds and Bartram, 2016).

These influences are primarily apparent through powers on the part of sustainability (Tian et al., 2022), resource shortage (De Laurentiis, Hunt, and Rogers, 2016), disturbances of networks between subdivisions and interdisciplinary (Cai et al., 2018), and the strengthening of the nexus in cities. The concept of safety within the nexus is generally robust, even though it is progressively observed as contentious. The prevailing safety narrative has been condemned for its neoliberal capitalist framing and its reductionist method (Middleton et al., 2015), which prioritises economic output over personal, administrations, or countries (Liebenguth, 2020), while overseeing the practical realities of the environment and related developments (De Grenade et al., 2016). This method shifts experts' focus from epicentre organisations to private institutions as the main administrators of nexus governance and security while also promoting educational solutions for hazard control (Liebenguth, 2020).

2.5.1 Nexus Activities and Their Timelines

Over the years, several research endeavours, conferences, and projects focused on the nexus concept have been organised globally. For instance, in 1983, the United Nations University (UNU) initiated the Food-Energy Nexus Agenda, identifying the complicated relationship between food and energy (Sachs and Silk, 1990). During the subsequent year, the "Food, Energy, and Ecosystems" conference took place in Brasilia, Brazil, organised by the organisation (UNU). In 1986, the UNU once more organised another International Symposium on "the Food-Energy Nexus and Ecosystems" in New Delhi, India. Simultaneously, the western United States, in the mid-1980s, investigated the relationship between electricity and water. As the 1990s began, the World Bank introduced the term "nexus" to connect water, food, and trade (McCalla, 1997). Starting from the middle to late 1990s through the beginning of the 2000s, the Columbia Water Center of the Earth Institute at Columbia University extensively investigated the water-energy-agriculture nexus in India. Afterwards, Scott, C.A. utilised the knowledge of the electricity-for-water nexus in Mexico (2011). Additionally, the discussion on "virtual water" and "water footprints" improved the nexus conversation (Allan, 2003). The demand to include energy as an essential element of the nexus was broadly recognised by researchers and experts worldwide, mostly during the Kyoto World Water Forum 2003 (Hussey and Pittock, 2012).

In 2011, an important landmark arose at the Bonn Nexus Conference, where the importance of the three nexus pillars—water, energy, and food—was formally publicised following the ideas of the Rio Plus 20, which underscored the idea of a "green economy." Moreover, following the Bonn Nexus Conference, the German Federal Government established the Water, Energy, and Food Security NEXUS Resource Platform. In 2012, the United Nations University Institute for Integrated Management of Material Fluxes and Resources (UNU-FLORES) in Dresden was also recognised, concentrating on the "integrated management of environmental resources: water, waste, and soil." In 2013, the United Nations Economic and Social Commission for Asia and the Pacific (UN-ESCAP) organised "The Status of the Water-Food-Energy Nexus in Asia and the Pacific." Concurrently, the FAO released "The Innovative Accounting Framework for the Food-Energy-Water Nexus: Application of the Multi-Scale Integrated Analysis of Societal and Ecosystem Metabolism (MuSIASEM) method to three case studies." Furthermore, the International Institute for Sustainable Development (IISD) described "The Water-Energy-Food Security Nexus: Towards a practical planning and decision-support framework for landscape investment and risk management" in 2013.

2.6 DESCRIBING THE FEW NEXUS PLANNING

The concept of the FEW nexus involves the combined management of food, energy, and water resources to improve their safety in times of challenges such as environmental degradation, constant depletion, and climate change. It acknowledges that any improvements or impediments in one of the three sectors will impact the others (FAO, 2016; Mahlknecht and González-Bravo, 2018; Nhamo, Ndlela, et al., 2020). This method shows a transformative and recurring model that broadly directs the organisation of these complicatedly related resources and systems, targeting sustainable development (Hoff, 2011; FAO, 2016). A cross-sectoral model is crucial for recognising portions of instant intervention, finding possible synergies, and identifying trade-offs that require pressing intervention (Mahlknecht and González-Bravo, 2018; Nhamo, Mabhaudhi et al., 2020; Naidoo et al., 2021).

Even though the FEW nexus is widely recognised, other nexus categories have been known, including the water-health-environment-nutrition (WHEN) nexus and the urban nexus, among others (Lehmann, 2018). The occurrence of various nexuses highlights the necessity to expansively evaluate and manage integrated sectors without selecting any one explicitly (Nhamo, Ndlela et al., 2020). Consequently, the FEW nexus remains the only widely recognised nexus among others (Willis, 2015).

With a different type of nexus, the idea has advanced beyond the thin sector of the FEW nexus to include a broader approach to nexus planning (Mabhaudhi et al., 2021). The FEW nexus examines the synergies and trade-offs between several areas, promoting productivity, resource mobilisation, and avenues for clear policymaking. However, this recent advancement has dismissed the concept that any nexus category is connected uniquely to the FEW nexus. The more comprehensive term "nexus planning" highlights the effort to provide integrated solutions for different yet interconnected components (Naidoo et al., 2021). Its transmutative and multifaceted nature eases the navigation of synergies and trade-offs while guiding the development and execution of coherent policies to steer the change towards robust socio-ecological and sustainable systems that ultimately enhance environmental outcomes and human well-being (Mabhaudhi et al., 2019).

2.6.1 Sustainable Food Resources

Specifically, goal 2 of the SDGs, which focuses on improving diets, promoting sustainable agriculture, ensuring food security, and ending hunger, has incorporated an evaluation of production sustainability (United Nations General Assembly, 2015). Goal 2 involves many components, including feeding people, food uncertainty, income, sustainability, and production. It equally considers the security of plant and animal genomes, governmental efforts on food security, and the operational functioning of food markets (United Nations General Assembly, 2015). These components serve as measures for evaluating the overall sustainability of food security.

Moreover, Target 2.3 of SDG aims to

> double agricultural output and the earnings of small-scale food producers, mainly women, native societies, domestic farmers, pastoralists, and fishers, by 2030. This comprises guaranteeing safety and reasonable means to land, other fruitful resources, knowledge, economic services, markets, and prospects for value addition and non-farm service.

This target is assessed by an output parameter. On the other hand, SDG Target 2.4 directly evaluates agriculture sustainability. It is designed to "guarantee sustainable food production systems and execute strong agricultural services by 2030. These services should grow output, sustain ecosystems, build capacity for climate change acclimatisation, severe weather, environmental disasters, flooding, address drought, and steadily improve land and soil quality."

Recently, nexus planning has evolved as a vital choice support means for widely resolving the intricate and interconnected essential resources. The three critical resources—food, energy, and water—reinforce the economic, environmental surfaces and social sustainability (Figure 2.1). This context provides an avenue to grasp and systematically examine socio-ecological networks and to articulate interconnected policies for a better united cross-sectoral resource management. This establishes the essential aspect of the idea, as it directs the complementary performance between synergies and trade-offs in an integrated way (Nhamo, Mabhaudhi et al., 2020).

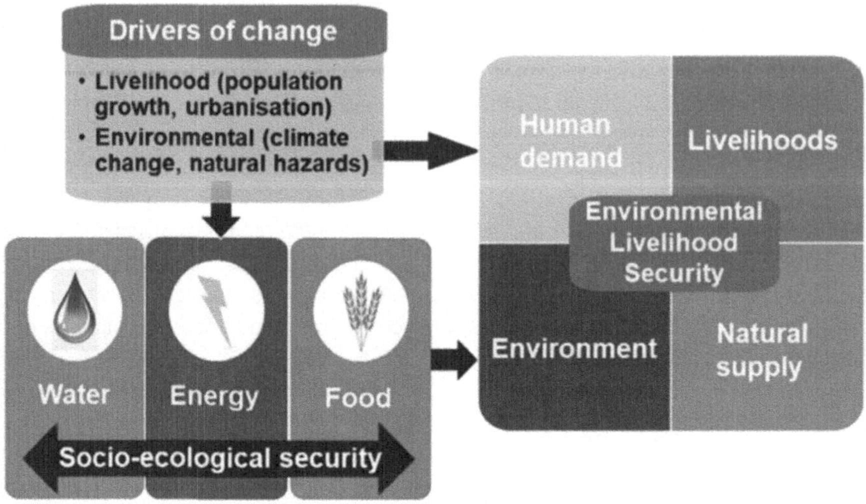

FIGURE 2.1 Drivers of transformation affecting the socio-ecological systems and how they are connected to the WEF nexus (Liphadzi et al., 2021).

The FEW nexus is a multifaceted and transformational method that safeguards these vital resources (Luxon and Pius, 2012). Thus, the FEW nexus represents a three-dimensional basis that serves as a systematic means, a conceptual and discourse framework (Luxon and Pius, 2012). Analytically, it steadily uses quantitative and qualitative approaches to comprehend the complex relationships among FEW resources. It defines and disintegrates the problems within the interconnectivity of the FEW sectors, encouraging sustainability. As a discourse, it is a problem-solving means to improve cross-sectoral relationships (Luxon and Pius, 2012). Furthermore, nexus planning is significant for evaluating the productivity of the FEW elements and the SDGs (Stephan et al., 2018). These basic features of the FEW nexus have played a crucial role in enriching it into a critical agenda in decision-making, mainly within sustainability domains (Leck et al., 2015).

2.6.2 Sustainable Energy Resources

Sustainable energy, which includes energy efficiency, utilisation, and accessibility of renewable resources, is crucial in promoting all SDGs. Globally, considerable efforts have been executed in search of SDG 7, accompanied by an alternative shift towards a sustainable energy supply. In developing countries, a constant improvement in the growing electricity supply has been observed, rising between 83% in 2010 and 87% in 2017 worldwide. The global use of renewable resources and energy efficiency has increased from 8.6% in 2010 to 10.2% in 2016. Nevertheless, there is an urgent demand to expand access to clean and secure technologies and cooking fuels for three billion people worldwide, as progress in this aspect has been slow, with only about 0.5% annual growth from 2010 to 2017 (UN-ESCAP et al., 2017). There is an

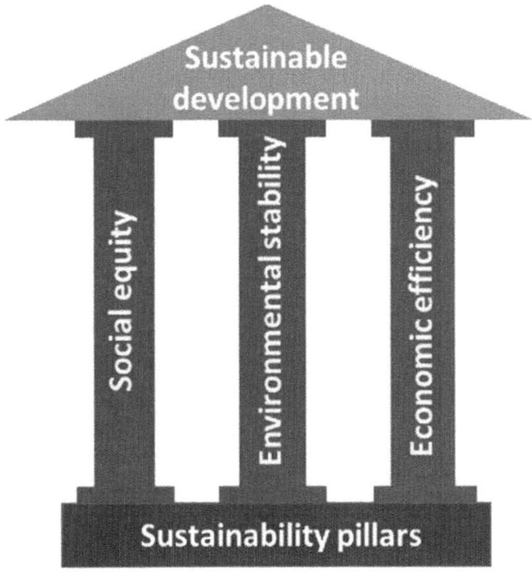

FIGURE 2.2 The concept of energy sustainable development (Liphadzi et al., 2021).

urgent need to increase power supply, especially in sub-Saharan Africa, because only 44% of the population has access to power supply (Fritz et al., 2019).

Conversely, the UN Report entitled "The Future We Want," which serves as the basis for the SDGs, describes sustainable development as reducing disparities, fostering inclusive and equal economic growth, improving the standard of living, increasing opportunities for all, promoting fair social growth and inclusion, and encouraging wide-ranging opportunities. Also incorporated in this is the sustainable integration of natural resources and ecosystems that sustain, among other things, economic, social, and human development. This includes maintaining, transforming, and restoring ecosystems as well as enhancing robustness in the face of rising challenges (United Nations General Assembly, 2015). The concept of energy sustainable development centres that focus on the ability to maintain balances among three vital areas: social fairness, environmental flexibility, and economic efficacy is presented in Figure 2.2.

2.6.3 Sustainable Use of Water-related Ecosystems

Target 6.6 of Sustainable Development concentrates on defending and re-establishing water-connected ecosystems in various environments such as forests, mountains, rivers, wetlands, lakes, and aquifers. While Goal 6 mainly focuses on ensuring the availability of clean water and sanitation services, Target 6.6 emphasises the crucial function of ecosystems in providing these services to society. Therefore, controlling and efficiently managing these ecosystems is vital to make sustainability an imaginable possibility. Preferably, the information for Target 6.6 should incorporate data on the environmental area of water-linked ecosystems, the water volumes they embrace, the quality of water around them, and the complete living standard the ecosystems incorporate. Sustainable development involves growth that does not threaten the life sustaining systems of the Earth's surface (Apeh and Nwulu, 2025). Unfortunately, accomplishing this has become a challenging task due to human interference, which has changed environments vital to human existence, resulting in various challenges such as climate change and unparalleled socio-ecological dynamics that threaten human well-being. This is demonstrated by the development of new infectious diseases such as COVID-19, HIV/AIDS, and Ebola, which have caused the loss of millions of lives.

2.7 TECHNOLOGICAL DEVELOPMENTS OF FEW SYSTEMS IN WATER RESOURCES

FEW-related infrastructure design and water technologies are swiftly evolving, leading to novel synergies and enhanced total system efficiency. Given its essential role in meeting the water needs of the food and energy sectors, water infrastructure is integrally appropriate to the FEW nexus. Nevertheless, it is regularly observed that water technologies and infrastructure design disregard the mutuality within the FEW sectors (Webber, 2015). There are roughly 3,700 large dams currently in the planning or designing stage worldwide, presenting a vital opportunity to integrate broader FEW system products into these novel dams' preparation, designing, and functioning (Zarfl et al., 2015). The emphasis of novel water infrastructure and technologies should be on exploiting all of the FEW benefits, including investors from different ranges, such as environmental considerations.

2.7.1 Technology Innovations

Several infrastructure designs and water technologies are great avenues to decrease trade-offs, improve efficiencies, or enhance consistency in FEW systems. For example, progress in both multi-phase reverse osmosis and flash distillation technologies has improved the energy efficiency of desalination plants, thus decreasing the water versus energy trade-off (Khawaji, Kutubkhanah and Wie, 2008). The advancement of drought-endurance crops (Tester and Langridge, 2010) can alleviate water scarcity susceptibility and store energy for irrigation systems. There are other instances, such as the utilisation of saline water for irrigating some salt-endurance crops (Rozema and Flowers, 2008), the recycling of effluent after profitable treatment (Becerra-Castro et al., 2015), and the application of waste food for bioenergy generation by low environmental effect technologies (Kiran et al., 2014).

An alternative pathway to achieving FEW savings comprises the application of waste as an appreciated resource, a programme recently sponsored by wastewater technologists. Specifically, domesticated water is sent to treatment plants for thermoelectric power production or used to irrigate biofuel and food crops (Dong et al., 2016). Wastewater treatment systems have proved the financial possibility of improving thermal energy for heating, incinerating organic compounds for energy generation, and applying nutrients as crop fertilisers (Hering et al., 2013). Through advanced resource retrieval technologies, the downstream water and environmental operators were assisted with better water quality. Moreover, farmers who used captured nitrogen and phosphorus from wastewater on their fields experienced improved crop outputs and reduced fertiliser costs.

However, water scientists should remain vigilant, as demonstrated by the FEW-related cases observed by Rajagopal (2008). In this case, societal advantages suddenly reduced as "wasteland" was transformed into agricultural land for expanding biofuel production. What designers had regarded as a "wasteland" was essentially an appreciated resource for a disregarded community, helping as a means of fuel and food for local penurious societies (Rajagopal, 2008). As food, energy, and water systems continue to integrate, new prospects may develop for water scientists to use "waste" from these systems or transmit it to places where it can be used as an appreciated resource. Though partially executed, some research studies and plans have continued the assessment of infrastructure and water technology comprising the impacts within a FEW system. To this end, Villarroel Walker et al. (2014) conducted studies on the impacts of four water-sector technologies on the "metabolism" of the city of London—the city's consumption and yield of nutrients, water, and energy (Walker et al., 2014).

Similarly, there is proof that regionalised hydropower systems and small-scale water storage may efficiently stabilise the numerous purposes of food, water, and energy safety in emerging nations (van der Zaag and Gupta, 2008). Regionalised methods have provided greater access to water and energy resources for food production among smallholder farmers who are in charge of a significant part of the food supply in impoverished countries (Stevens, 2015). Moreover, the more robust integration between resident investors and water infrastructure may produce extra FEW synergies, including using river sediments rich in nutrients to cropland and enhancing food production while at the same time extending the life of the dam. Authorities in water resource systems should gradually deliberate on the integrated processes and systems of the FEW nexus when assessing technology and infrastructure. The increasing competencies of technologies such as geographic information systems (GIS), smartphone applications, remote sensing, and affordable sensors can supply vital data related to FEW systems, possibly by assisting water researchers in conducting more integrated investigations. For instance, data retrieved from old water infrastructure reconstructed with smart sensors can provide better adaptable functional measures and increase decision support conception, eventually leading to more effective utilisation of FEW resources and a more environmentally sustainable process.

However, fundamental questions must be addressed when developing these innovative data devices: how can diverse and disjointed users supervise and assess resource usage? How can information about stakeholder behaviours, institutions, and human processes in various areas effectively combine with ecosystem and physical data? What data is required to comprehend the temporal and spatial growth of FEW systems? As responses to these questions are formulated and as more excellent sensing competencies are connected, water and FEW system experts will be better prepared to comprehend the difficulty of these systems, providing directions for prospective infrastructural and technological investments.

2.7.2 Systems Design

The FEW nexus is determined to allow water scientists to self-determine and integrate current infrastructures and technologies in new methods for better total FEW systems (Garcia and You, 2016). Ancient water usage, infrastructure, and technology can be reformed to achieve synergies within the broader FEW systems. The combination of interconnection on FEW processes in design creates prospects to progress resource application efficiency and attain synergistic societal support. One of the approaches includes minimising the energy required to provide clean water through the combination of natural watershed management and engineered water treatment. For instance, biofuel generation from persistent herbaceous crops in peripheral agricultural land can decrease residue and excess nutrients, enhance water quality, and decrease the requirement for energy-concentrated water therapy before human intake. On the other hand, several energy sources could also be doubled for generation purposes. Because of varying weather conditions as well as the timing of energy demands, solar, hydropower, and wind energy can be combined to increase energy production dependability.

As an illustration, General Electric has recently established a plan using large-scale wind turbines to pump water onto a hill for storage purposes. At the time when electricity is required, the water being stored is released downhill and passed through a hydropower producer. This system employs water as a battery to store surplus power produced by the wind turbines. This idea is an expansion of the "pump and hydropower generation" technique, where water is driven to a higher storage tank during off-peak hours (when energy need is lower) and then released (with its energy harnessed) during peak hours when both electricity and water demand are higher (Grumet, 2016). Besides, approaching challenges usually considered as "water-related issues" (such as floods) from the viewpoint of FEW systems will transform our approach to handling, designing, and articulating rules for water-related challenges. As an illustration, erosion and coastal flooding are usually moderated by structural (such as dykes and barriers) or strategic ways (such as flood protection); so far, it is difficult to understand the appearance of FEW in the framework of flood resilience. One instance could be observed in small island countries, where the restoration of mangrove environments improves and guards local inhabitants from coastal flooding and erosion, enhances water quality, and generates food with negligible interruption to energy and water supplies (Biggs et al., 2015). As shown in the above illustrations, several significant synergies are only available to system engineers if food, energy, and water systems are examined and achieved jointly. Water researchers and experts must propose untraditional methods to address the increasingly scarce FEW resources.

2.7.3 Externalities of Technology Development

Modern systems and technologies always involve great assets to advance and execute, and such assets may be misused or even generate extra challenges if they are handled with inadequate knowledge of FEW system intricacies (Webber, 2015). Thus, systems specialists in water and other fields have a great responsibility to guide infrastructure development and technology growth with expanding knowledge of FEW system intricacies. Cases of technology misuse causing unplanned costs and shocks are plenty. In water research, a typical instance is the rise of water intake due to enhanced irrigation technologies (Dogo et al., 2019). This case, termed Jevon's Paradox, also happens in the food and energy areas (Freeman, Yearworth and Preist, 2016). Secondly, water reuse or recycling plans can increase water utilisation productivity but have unplanned effects across the nexus. For example, frequent reuse and recycling of irrigation excess adversely impacted water quality in Egypt's irrigation system. The water became salty and contaminated as it was reprocessed, decreasing crop productivity.

Moreover, the upstream–downstream scenarios of water reprocessing generated challenges of inequality for Egyptian farmers (Barnes, 2014). Hence, two spontaneous remedies to water resource shortage—enhanced productivity and resource recycling—must not be followed quickly but through cautious deliberation of their wider effects within the FEW systems and society (Apeh and Nwulu, 2024). These illustrations equally motivate emerging water management approaches. As a result of the complicated nature of FEW systems, even accurately crafted plans can produce unexpected results. Adaptive management methods enable water directors to modify their policies based on recently developed knowledge (Pahl-Wostl, 2007). For instance, in 1980, with the passing of the Northwest Power Act, which authorised the sustainability of hydropower expansion with the preservation of fish populations, the Columbia River Basin Fish Wildlife Program (CRBFWP) established an adaptive control approach.

The CRBFWP instantly introduced alleviation efforts while establishing devices to evaluate the effects of habitat improvements on salmon spawning rates as they were executed. Consequently, the CRBFWP has sustained an ongoing learning process and has adjusted its management policies. Generally, infrastructure design and technology utilisation focused on considering only input–output connections or other separate procedures may be mistaken concerning particular general system features (Kumar, 2015). As knowledge of FEW system complications advances, unplanned consequences of technology and design can be averted more easily.

2.8 NEXUS REGIONS

Figure 2.3 presents the various nexus projects and their distribution in several world regions. These areas were distinguished into Europe, Oceania, South America, Asia, North America, the Middle East, and Africa. In this category, the Middle East is considered different from Asia because of a critical number of enduring nexus plans. If a nexus plan is executed globally, it is calculated as a project happening everywhere. In this evaluation, it is noticed that projects numbering up to six were executed globally: one project each for water–energy and water–energy and food, two projects for water–food, and two for climate-connected. Oceania and North America had a propensity to concentrate on an exact nexus type, climate-linked (43%) and water–energy (46%), while Africa had less emphasis on water–energy (7%). However, the other places had comparatively stable attention in each nexus type.

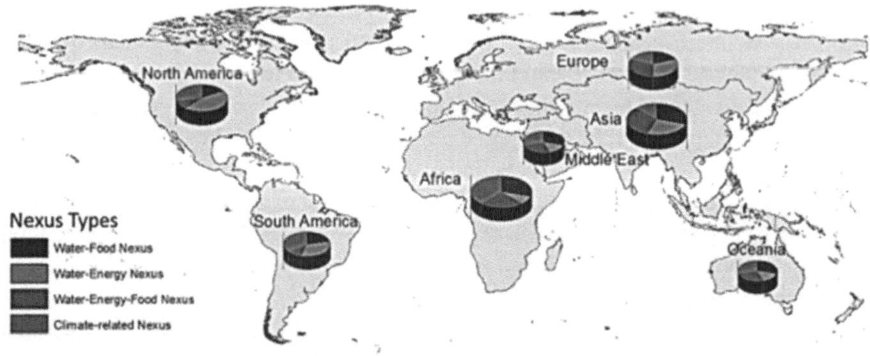

FIGURE 2.3 Representation of nexus types in various regions.

2.9 TRADE-OFFS, SYNERGIES, AND SYSTEM EFFICIENCY

Nexus-oriented thinking demands a change from emphasising exploiting generation to increasing system efficiencies, harnessing synergies, and expertly steering trade-offs. The difficulty and different water applications have long required this holistic viewpoint, preceding the importance of the FEW nexus in scientific discussions. While the past efforts of water scientists and present can contribute to the FEW nexus studies, the FEW agenda possibly offers a pathway for bigger application of our efforts by including and engaging persons who have earlier avoided a water-centric method. In the water systems domain, professionals, policymakers, and experts have long contended with balancing considerations between water quality and quantity, listing upstream receivers versus downstream receivers, and considering environmental well-being against economic productivity, among other factors. In the background of FEW, these trade-offs become more complex as different sectors and investors are engaged. To expansively understand water resources, we must utilise nexus thinking. For instance, in central Illinois, Housh et al. (2015) experimented by implementing a second-generation biofuel crop like Miscanthus in a watershed (Housh et al., 2015). This aim is to increase the efficiency in biofuel production and decrease nutrient release, although at the expense of improved cost and higher freshwater utilisation compared to the present application of corn as feedstock. The FEW nexus method authorises surpassing the trade-off of water quality against quantity and investigates the fundamental devices to navigate the system products. The promoters of this system include farm organisations, energy and food markets, developing energy technologies, transportation infrastructure, and rates of feedstock conversion. Spreading the limits of our examination will disclose earlier concealed and ineffectively understood trade-offs. Resolving these complex trade-offs demands higher scientific support for understanding procedures, interventions, and results and encourages institutional backing to harmonise the advantages for different groups of investors.

Furthermore, FEW systems integration can produce positive synergies, contributing to a remedy of long-standing problems within the water sector; for instance, in areas categorised as arid or semi-arid weather, where there is a shortage of water resources and regular low quality (such as saline groundwater and wastewater) and a region of plentiful renewable energy sources including wind and solar power. These energy sources can be utilised to desalt groundwater and treat wastewater. The treated resources can then be utilised for high-value crop irrigation through vertical agriculture—a modern method that includes stacked layers of cultivated land within a building. This connection of energy and water resources for the production of food and storage can assist in many functions, such as (1) improving the applications of renewable energies, as capacities of windmills and solar systems remain underused in many places; (2) expanding food production and limiting food wastage; (3) converting marginal water sources (e.g., wastewater and saline water) into beneficial assets. The possible synergies in growing from integrated FEW systems produce beneficial understandings for scientists concentrating on FEW and engineers accounted for when designing FEW systems to fit in local resources and socio-economic situations. However, solving critical issues such as environmental effects, social consequences, and cost-effectiveness is essential.

However, the estimated synergies of integrated FEW systems may deteriorate or even be substituted by trade-offs because of fluctuations in social and environmental circumstances. Multi-purpose reservoirs within water resources serve as a critical example. Dams are engineered to achieve several purposes, such as irrigation, hydropower, public water supply and fishing. During the design stage, ideas like crop irrigation and energy generation are equalised to meet the instant requirements of those needing the dam's water discharges. Nevertheless, long-term damming of upstream waters, variations in land use, and changes in climate patterns have interrupted hydrological patterns and the timing of water necessity for both food production and hydropower generations (Zeng et al., 2017), thereby affecting the dam's working efficiency (Zeng et al., 2017). Furthermore, variations in social values about the fish and environmental habitats have raised questions about past water distributions (Marston and Cai, 2016).

Change in society and the natural world is inevitable and seems to be increasing. Collaboratively, FEW researchers can boost the flexibility of the FEW sectors to potential variations by deploying flexible infrastructure and applying additional resources from every segment. An instance of this is the practice of managed aquifer recharge (MAR) and conjunctive water use (CU), in which surplus water is reserved in exhausted aquifers for future withdrawal to irrigate biofuel crops or food during droughts (Scanlon et al., 2016). More technological developments, combined with forward-thinking strategies, can produce extra FEW synergies and strengthen the flexibility of each sector to variations in natural and human systems. Improving water utilisation efficiency within the food and energy sectors (the two major water consumers worldwide) has been a long-lasting search in technology communities and water science.

Nevertheless, it is imperative to highlight that expanding water utilisation efficiency does not certainly enhance system productivity when considering wider or various FEW purposes. Even in the domain of the FEW nexus, the essential concept of efficiency requires reconsideration. For example, closed-loop thermoelectric cooling is regarded as more efficient than open-loop cooling as it extracts less water by nearly 95%. Nevertheless, closed-loop cooling can consume approximately 60% more water than open-loop cooling (DeNooyer et al., 2016). Enhancing irrigation productivity might decrease the amount of water used for crops, but extra inputs in energy are essential to force these extra water-efficient irrigation systems. Besides, efficiency measurements are highly dependent on the limits of the examination. Improvements in the efficiency of water at the localised level may employ a slight impact on basin efficiency (described as the ratio of helpful water consumption over total water consumption within a basin) since return flows from "inefficient" water users can be exploited downstream for hydro-energy generation, food production, and other intentions (Cai, Rosegrant and Ringler, 2003).

2.10 CONCLUSION

The chapter describes the possibility of harnessing natural resources in biodiverse countries such as Mexico, China, and various countries in Asia and Africa to reduce poverty and encourage sustainable development. It highlights the importance of efficiently managing resources to generate job opportunities and improve social communications and democracy. Integrating ecological parameters with economic ones is emphasised as a critical approach for attaining these purposes. Motivations for sustainable resource management are vital for energy savings, decreased pollution, and financial benefits. However, challenges, such as inequality in access to green technology and patents held by transnational establishments, worsen global inequalities. The chapter highlights the importance of allowing local communities, mainly indigenous groups, to be involved in environmental conservation drives. Global environmental conferences and enterprises have addressed sustainable development, biodiversity, and climate change challenges, stressing the necessity for harmonised action. Human ecology is introduced as an agenda for encouraging welfare while addressing cultural biases and societal threats. The chapter then explores the evolution of the nexus notion, which integrates food, energy, and water resources with governance and socio-economic features. Numerous nexus agendas are discussed, emphasising the necessity for interdisciplinary methods to address complex challenges. The chapter also summarises historical milestones and study efforts connected to the nexus idea, tracing its development from the 1980s to the present. Furthermore, the chapter puts forward the FEW nexus planning methods, which intend to manage these scarce resources sustainably and integrate them. It underlines the interrelation of food, energy, and water systems and their effect on socio-economic and environmental sustainability. The FEW nexus is depicted as a multi-faceted device for understanding and addressing intricate challenges that are significant for attaining SDGs. Generally, the chapter emphasises the importance of holistic and collaborative strategies for resource management and sustainable development, highlighting the necessity for integrated plans and approaches to address interconnected challenges efficiently.

REFERENCES

Albrecht, T. R., Crootof, A. and Scott, C. A. (2018) 'The water-energy-food nexus: A systematic review of methods for nexus assessment'. *Environmental Research Letters*, 13(4), p. 43002.

Allan, J. A. (2003) 'Virtual water – the water, food, and trade nexus. Useful concept or misleading metaphor?' *Water International*, 28(1), pp. 106–113.

Apeh, O. O., Meyer, E. L. and Overen, O. K. (2022) 'Contributions of solar photovoltaic systems to environmental and socioeconomic aspects of national development—A review'. *Energies*, 15(16), p. 5963.

Apeh, O. O. and Nwulu, N. I., (2024). The water-energy-food-ecosystem nexus scenario in Africa: Perspective and policy implementations. *Energy Reports*, 11, pp. 5947–5962.

Apeh, O. O., & Nwulu, N. I. (2025). Enhancing transparency and efficiency in green energy management through blockchain: A comprehensive bibliometric analysis. *Energy Nexus*, 100405.

Barnes, J. (2014) 'Mixing waters: The reuse of agricultural drainage water in Egypt'. *Geoforum*, 57, pp. 181–191.

Becerra-Castro, C. et al. (2015) 'Wastewater reuse in irrigation: A microbiological perspective on implications in soil fertility and human and environmental health, *Environment International*, 75, pp. 117–135.

Biggs, E. M. et al. (2015) 'Sustainable development and the water–energy–food nexus: A perspective on livelihoods'. *Environmental Science & Policy*, 54, pp. 389–397.

Bizikova, L. et al. (2013) *The water-energy-food security nexus: Towards a practical planning and decision-support framework for landscape investment and risk management.* International Institute for Sustainable Development, Winnipeg.

Bryceson, D. and Vuorela, U. (2020) *The transnational family: New European frontiers and global networks*. Routledge.

Cai, X., Rosegrant, M. W. and Ringler, C. (2003) 'Physical and economic efficiency of water use in the river basin: Implications for efficient water management'. *Water Resources Research*, 39(1) pp. 11.

Cai, X. et al. (2018) 'Understanding and managing the food-energy-water nexus – Opportunities for water resources research, *Advances in Water Resources*, 111, pp. 259–273.

Cairns, R. and Krzywoszynska, A. (2016) 'Anatomy of a buzzword: The emergence of "the water-energy-food nexus" in UK natural resource debates'. *Environmental Science & Policy*, 64, pp. 164–170.

Carmona-Moreno, C., Dondeynaz, C. and Biedler, M. (2019) *Position paper on water, energy, food and ecosystems (WEFE) nexus and sustainable development goals (SDGs).* Publications Office of the European Union.

Chen, S. et al. (2022) 'Updating global urbanization projections under the shared socioeconomic pathways'. *Scientific Data*, 9(1), p. 137.

Convention on Biological Diversity (CBD) (2020) *CBD – Global biodiversity outlook 5 – Full report*. www.emdashdesign.ca

De Grenade, R. et al. (2016) 'The nexus: Reconsidering environmental security and adaptive capacity. *Current Opinion in Environmental Sustainability*, 21, pp. 15–21.

De Laurentiis, V., Hunt, D. V. L. and Rogers, C. D. F. (2016) 'Overcoming food security. Challenges within an energy/water/food nexus (EWFN) approach'. *Sustainability*, 8(1), p. 95.

DeNooyer, T. A. et al. (2016) 'Integrating water resources and power generation: The energy–water nexus in Illinois'. *Applied Energy*, 162, pp. 363–371.

Distefano, T. et al. (2022) 'Sub-national water–food–labour nexus in Colombia'. *Journal of Cleaner Production*, 335, p. 130138.

Divides, K. (2010) *World social science report*. UNESCO.

Dodds, F. and Bartram, J. (2016) *The water, food, energy and climate nexus: Challenges and an agenda for action*. Routledge.

Dogo, E. M. et al. (2019) 'Blockchain and internet of things-based technologies for intelligent water management system'. *Artificial intelligence in IoT* (pp. 129–150). Springer.

Dong, S. et al. (2016) 'Comparative mammalian cell cytotoxicity of wastewaters for Agricultural reuse after ozonation'. *Environmental Science & Technology*, 50(21), pp. 11752–11759.

Elhacham, E. et al. (2020) 'Global human-made mass exceeds all living biomass'. *Nature*, 588(7838), pp. 442–444.

Endo, A. et al. (2017) 'A review of the current state of research on the water, energy, and food nexus'. *Journal of Hydrology: Regional Studies*, 11, pp. 20–30.

Fanzo, J. et al. (2018) *2018 global nutrition report: Shining a light to spur action on nutrition* Development Initiatives.

FAO (2016) 'Agriculture organization, 2014'. In *Livestock primary*. Food and Agriculture Organization of the United Nations.

Freeman, R., Yearworth, M. and Preist, C. (2016) 'Revisiting Jevons' paradox with system dynamics: Systemic causes and potential cures'. *Journal of Industrial Ecology*, 20(2), pp. 341–353.

Friedman, R. S. et al. (2018) 'How just and just how? A systematic review of social equity in conservation research'. *Environmental Research Letters*, 13(5), p. 53001.

Fritz, S. et al. (2019) 'Citizen science and the United Nations sustainable development goals'. *Nature Sustainability*, 2(10), pp. 922–930.

Garcia, D. J. and You, F. (2016) 'The water-energy-food nexus and process systems engineering: A new focus'. *Computers & Chemical Engineering*, 91, pp. 49–67.

Glass, L.-M. and Newig, J. (2019) 'Governance for achieving sustainable development goals: How important are participation, policy coherence, reflexivity, adaptation and democratic institutions?' *Earth System Governance*, 2, p. 100031.

Grumet, T. (2016) 'This unique combo of wind and hydro power could revolutionize renewable energy'. *GE Reports*, 18.

Handl, G. (2012) 'Declaration of the United Nations conference on the human environment (Stockholm Declaration), 1972 and the Rio declaration on environment and development, 1992'. *United Nations Audiovisual Library of International Law*, 11(6) pp. 1–11.

Hejnowicz, A. P. et al. (2022) 'Appraising the water-energy-food nexus from a sustainable development perspective: A maturing paradigm?' *Earth's Future*, 10(12), p. e2021EF002622.

Hendriks, C. M. (2006) 'Integrated deliberation: Reconciling civil society's dual role in deliberative democracy'. *Political Studies*, 54(3), pp. 486–508.

Hering, J. G. et al. (2013) *A changing framework for urban water systems*. ACS Publications.

Hoff, H. (2011) *Understanding the nexus*. Stockholm Environment Institute (SEI).

Housh, M. et al. (2015) 'Managing multiple mandates: A system of systems model to analyze Strategies for producing cellulosic ethanol and reducing riverine nitrate loads in the Upper Mississippi River Basin'. *Environmental Science & Technology*, 49(19), pp. 11932–11940.

Hussey, K. and Pittock, J. (2012) 'The energy-water nexus: Managing the links between energy and water for a sustainable future'. *Ecology and Society*, 17(1) p. 31.

IEA. (2021) *The role of critical minerals in clean energy transitions*. IEA Paris.

Jamison, A. and Rohracher, H. (2001) *Introduction: Technology studies and sustainable Development*. Taylor & Francis.

Khawaji, A. D., Kutubkhanah, I. K. and Wie, J.-M. (2008) 'Advances in seawater Desalination technologies'. *Desalination*, 221(1–3), pp. 47–69.

Kiran, E. U. et al. (2014) 'Bioconversion of food waste to energy: A review'. *Fuel*, 134, pp. 389–399.

Kumar, P. (2015) 'Hydrocomplexity: Addressing water security and emergent environmental risks'. *Water Resources Research*, 51(7), pp. 5827–5838.

Lade, S. J. et al. (2020) 'Human impacts on planetary boundaries amplified by Earth system interactions'. *Nature Sustainability*, 3(2), pp. 119–128.

Laspidou, C. S., Mellios, N. and Kofinas, D. (2019) 'Towards ranking the water–energy–food–land use–climate nexus interlinkages for building a nexus conceptual model with a heuristic algorithm'. *Water*, 11(2), p. 306.

Lazaro, L. L. B. et al. (2022) 'Ten years of research on the water-energy-food nexus: An Analysis of topics such as evolution and frontiers in water. *Frontiers*, 4, p. 859891.

Lebel, L. and Lebel, B. (2018) 'Nexus narratives and resource insecurities in the Mekong Region'. *Environmental Science & Policy*, 90, pp. 164–172.
Leck, H. et al. (2015) 'Tracing the water–energy–food nexus: Description, theory and practice'. *Geography Compass*, 9(8), pp. 445–460.
Lehmann, S. (2018) 'Implementing the urban nexus approach for improved resource-the efficiency of developing cities in Southeast Asia'. *City, Culture and Society*, 13, pp. 46–56.
Lewis, S. L. and Maslin, M. A. (2018) *The human planet: How we created the Anthropocene*. Yale University Press.
Licha, I. (1996) *La investigación y las universidades latinoamericanas en el umbral del siglo XXI: Los desafíos de la globalización*. Udual.
Liebenguth, J. (2020) 'Conceptions of security in global environmental discourses: Exploring the water-energy-food security nexus'. *Critical Studies on Security*, 8(3), pp. 189–202.
Liphadzi, S. et al. (2021) 'The evolution of the water–energy–food nexus as a transformative approach for sustainable development in South Africa'. *The water–energy–food nexus: Concept and assessments* (pp. 35–67). Springer.
Lutz, W., Butz, W. P. and Samir, K. C. ed (2017) *World population & human capital in the twenty-first century: An overview*. Oxford University Press.
Luxon, N. and Pius, C. (2012) 'Climate change risk and vulnerability mapping and profiling at the local level using the household economy approach (HEA)'. *Journal of Earth Science and Climatic Change*, 3(123), p. 2.
Mabhaudhi, T. et al. (2019) 'The water–energy–food nexus as a tool to transform rural livelihoods and well-being in Southern Africa'. *International Journal of Environmental Research and Public Health*, 16(16), p. 2970.
Mabhaudhi, T. et al. (2021) 'Assessing progress towards sustainable development goals through nexus planning'. *Water*, 13(9), p. 1321.
Mahlknecht, J. and González-Bravo, R. (2018) 'Measuring the water-energy-food nexus: The case of Latin America and the Caribbean region'. *Energy Procedia*, 153, pp. 169–173.
Marston, L. and Cai, X. (2016) 'An overview of water reallocation and the barriers to its implementation'. *Wiley Interdisciplinary Reviews Water*, 3(5), pp. 658–677.
Masson-Delmotte, V. et al. (2021) 'Climate change 2021: The physical science basis'. In *Contribution of working group I to the sixth assessment report of the intergovernmental panel on climate change* (p. 2). Cambridge University Press Cambridge.
McCalla, A. (1997). The water, food, and trade nexus. Paper delivered at the *MENA-MED Conference, convened by the World Bank*, Marrakesh, Morocco, 3–6 September 1997.
Middleton, C. et al. (2015) 'The rise and implications of the water-energy-food nexus in Southeast Asia through an environmental justice lens'. *Water Alternatives*, 8(1) 627–654.
Naidoo, D. et al. (2021) 'Operationalising the water-energy-food nexus through the theory of change'. *Renewable and Sustainable Energy Reviews*, 149, p. 111416.
Nhamo, L., Mabhaudhi, T., et al. (2020) 'An integrative analytical model for the water-energy-food nexus: South Africa case study, *Environmental Science & Policy*, 109, pp. 15–24.
Nhamo, L., Ndlela, B., et al. (2020) 'The water-energy-food nexus as an adaptation strategy to achieve sustainable livelihoods at a local level, sustainability is needed' *Sustainability*. 12(20), p. 8582.
Oswald Spring, Ú. and Oswald Spring, Ú. (2019) 'Peace and sustainability in a globalised world'. In *Úrsula Oswald spring: Pioneer on gender, peace, development, environment, food and water: With a foreword by Birgit Dechmann* (pp. 201–223). Springer.
Pahl-Wostl, C. (2007) 'Transitions towards adaptive management of water facing climate and global change'. *Water Resources Management*, 21, pp. 49–62.
Pant, L. P., Adhikari, B. and Bhattarai, K. K. (2015) 'Adaptive transition for transformations to sustainability in developing countries'. *Current Opinion in Environmental Sustainability*, 14, pp. 206–212.

Rajagopal, D. (2008) 'Implications of India's biofuel policies for food, water and the poor'. *Water Policy*, 10(S1), pp. 95–106.

Ramos, E. et al. (2021) 'Capacity development and knowledge transfer on the climate, land, water and energy nexus'. In Brouwer, F. (Ed.) *Handbook on the water-energy-food nexus*. Edward Elgar Pub.

Ravetz, J. R. (2020) *Scientific knowledge and its social problems*. Routledge.

Ringler, C., Bhaduri, A. and Lawford, R. (2013) 'The nexus across water, energy, land and Food (WELF): Potential for improved resource use efficiency?' *Current Opinion in Environmental Sustainability*, 5(6), pp. 617–624.

Ron, I. (2011) 'Global citizenship. A path to building identity and community in a globalized world'. *The Global Citizen's Initiative*. http://www.theglobalcitizensinitiative.org

Rozema, J. and Flowers, T. (2008) 'Crops for a salinized world'. *Science*, 322(5907), pp. 1478–1480.

Sachs, I. and Silk, D. (1990) *Food and energy: Strategies for sustainable development*. United Nations University Press.

Scanlon, B. R. et al. (2016) 'Enhancing drought resilience with conjunctive use and managed. Aquifer recharge in California and Arizona'. *Environmental Research Letters*, 11(3), p. 35013.

Schlör, H., Venghaus, S. and Hake, J.-F. (2018) 'The FEW-nexus city index–Measuring urban resilience'. *Applied Energy*, 210, pp. 382–392.

Scott, C. A. (2011) 'The water-energy-climate nexus: Resources and policy outlook for aquifers in Mexico'. *Water Resources Research*, 47(6) W00L04.

Slorach, P. C. et al. (2020) 'Environmental sustainability in the food-energy-water-health Nexus: A new methodology and an application to food waste in a circular economy'. *Waste Management*, 113, pp. 359–368.

Stephan, R. M. et al. (2018) 'Water–energy–food nexus: A platform for implementing the sustainable development goals'. *Water International*, 43(3), pp. 472–479.

Stevens, L. (2015) *The energy| water| food Nexus at decentralized scales*. Practical Action Publishing.

Subedi, R., Karki, M. and Panday, D. (2020) 'Food system and water–energy–biodiversity nexus in Nepal: A review, *Agronomy*, 10(8), p. 1129.

Tester, M. and Langridge, P. (2010) 'Breeding technologies to increase crop production in a changing world'. *Science*, 327(5967), pp. 818–822.

Tian, P. et al. (2022) 'Water-energy-carbon nexus in China's intra and inter-regional trade'. *Science of the Total Environment*, 806, p. 150666.

UN-ESCAP et al. (2017) *World economic situation and prospects 2017*. United Nations.

United Nations General Assembly. (2015, September 25). Transforming our world: The 2030 Agenda for Sustainable Development (A/RES/70/1). https://www.un.org/ga/search/view_doc.asp?symbol=A/RES/70/1&Lang=E

Van der Zaag, P. and Gupta, J. (2008) 'Scale issues in the governance of water storage projects'. *Water Resources Research*, 44(10) W10417.

von Werlhof, C. (2008) 'The globalization of neoliberalism, its consequences, and some of its basic alternatives'. *Capitalism Nature Socialism*, 19(3), pp. 94–117.

Walker, R. V. et al. (2014) 'The energy-water-food nexus: Strategic analysis of technologies for transforming the urban metabolism'. *Journal of Environmental Management*, 141, pp. 104–115.

Water, U. N. (2018) *2018 UN world water development report, nature-based solutions for water*. UNESCO.

Webber, M. E. (2015) 'A puzzle for the planet'. *Scientific American*, 312(2), pp. 62–67.

WHO/UNICEF (2022) *Progress on drinking water, sanitation and hygiene in schools*. https://data.unicef.org/wp-content/uploads/2022/08/JMP-WASH-in-schools_2022.pdf

Williams, M. et al. (2016) 'The anthropocene: A conspicuous stratigraphical signal of anthropogenic changes in production and consumption across the biosphere'. *Earth's Future*, 4(3), pp. 34–53.

Willis, C. (2015) 'The contribution of cultural ecosystem services to understanding the tourism–nature–wellbeing nexus'. *Journal of Outdoor Recreation and Tourism*, 10, pp. 38–43.

Zarfl, C. et al. (2015) 'A global boom in hydropower dam construction'. *Aquatic Sciences*, 77, pp. 161–170.

Zeng, R. et al. (2017) 'Hydropower versus irrigation—An analysis of global patterns'. *Environmental Research Letters*, 12(3), p. 34006.

Zhao, R. et al. (2018) 'Impacts of water and land resources exploitation on agricultural carbon emissions: The water-land-energy-carbon nexus'. *Land Use Policy*, 72, pp. 480–492.

3 The Food-Energy-Water-Ecosystem Nexus Scenario in Africa

Perspective and Policy Implementations

3.1 INTRODUCTION

The continuous increase in demand for FEWE nexus resources in the African continent has drawn the attention of many researchers and stakeholders within the region, even though many places remain inaccessible (Dargin, Daher and Mohtar, 2019). The continent is mainly enriched with food, energy, and water-connected natural resources. Yet, most countries within the region cannot boost adequate, clean, and sustainable FEWE resources for their people. Currently, statistics show that virtually half of the people in the region are starving, about 20% are malnourished, 47% lack access to electricity, and 36% cannot provide basic amenities, including freshwater (Muhirwa et al., 2023). The application of vast quantities of FEWE resources while the basic facilities are still lacking within African countries remains a major challenge (Marti and Puertas, 2020; Lin, Wambersie and Wackernagel, 2021). A recent study revealed that ecological shortages and food, energy, and water disparities are massive in several African countries (Muhirwa et al., 2022). The challenges in the shortage of FEWE resources, such as financial water shortage, which influences virtually every African country, are recognised (Jemmali, 2018). This shortage can be caused by the failure to meet the resource demand because of inadequate assets in production, meagre facilities, insufficient human capacity, or inadequate organisation (Damkjaer and Taylor, 2017). Figure 3.1 shows the percentage of resources imported in several African countries.

The sustainability of natural resources mainly relies on goods and services obtained from its natural assets. Controlling these resources, particularly water, energy, food, and the ecosystem, creates a big vacuum for sustainable development. Presently, there is a considerable possibility for growth in the agri-energy area, recording only 15–30% of the prospective hydropower being generated (Danert, Adekile and Canuto, 2020). Besides, the vast irrigation potential of its 64 cross-border river basins is unsecured for food and nutritional safety. Nevertheless, water insecurity is indisputably a restrictive element, aggravated by multifaceted hydrology, rising demand, and climate change. Hence, evaluating the main rising necessities for resources in the African setting is crucial within the rapid strategies for

DOI: 10.1201/9781003633846-3

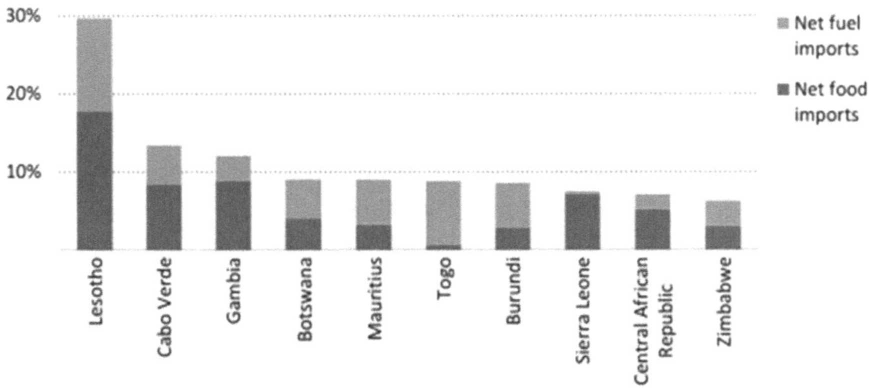

FIGURE 3.1 The percentage of net fuel and food imports in many African countries.

prolific water usage in the energy and agriculture sectors. The FEWE nexus approach is a strategy to resolve possible challenging demands for water for energy and agriculture while safeguarding desires for sustaining ecosystems. This must be done in an integrated, transboundary, and equitable way and by boosting cooperation across borders.

However, the tiny production of FEWE resources remains a significant challenge, as recognised by the Centre for SDGs in Africa, and it might persist in several countries of Africa (Twinoburyo et al., 2019). In recent studies, literature studies providing solutions to the SDGs connected to FEWE security are centred highly on improving accessibility of resources, consumption, and affordability, which are the SDG programme's essential targets (Belinskij, 2015). A few studies have addressed the challenges of ensuring a sustainable supply of FEWE resources to meet the growing food demand (IRENA 2015). The reliance on importing FEWE resources, such as agricultural facilities, by many African countries from outside the shore is quite high. For instance, in 2014, more than US$35 billion was used to purchase food resources, which is expected to rise and reach US$110 billion by 2025 (Union, 2014). However, several plans and strategies have been established under Africa's Programme Agenda 2063 to solve the challenges, including growing local generation, decreasing food imports by 50% by 2045, and reducing climate change's effect (Hirwa et al., 2021; Foong et al., 2022). Economic growth, accessible quality healthcare, and education boost society by promoting and assisting development and sustainability. These interconnected elements strengthen each other to allow and increase growth.

Moreover, apart from being an essential idea for relating societal influences and policymakers, Miller and Wyborn (2020) recognised that research-solving nexus investigations are not recognised in sustainability studies (Miller and Wyborn, 2020). It is also clear that FEWE nexus and SDG-connected study areas are still undiscovered and need extra quantification and study. Research by Ranzato et al. (2018) determined that studies on FEW nexus and waste services are essential for promoting sustainability and ensuring safety in the surrounding (Ranzato et al., 2018). Every piece of resource production has cooling systems, services, and other variables that should be measured in FEWE nexus resources. The study conducted by Muhirwa et al. (2022) revealed that the idea of FEW analysis is essential for designing FEW nexus and SDG related initiatives, particularly on the side of generation demand, which presently encounters serious issues in meeting demand (Muhirwa et al., 2022). Similarly, Muhirwa et al. (2022) recognised that the nexus idea allows the analysis of several countries into a system of networking (Muhirwa et al., 2023). In this work, researchers are mainly concerned with deliberating the system networking of food, energy, and water production through the lens of transboundary collaboration to advance the three areas' nexus efforts. The FEWE syntheses under nexus concepts are beneficial in helping investors recognise and relate multidimensional significances in every resource area's production, decrease resource excess compromises, and sustain the recognition of the SDGs (Putra, Pradhan and Kropp, 2020; Zhang et al., 2022). The current study on the FEWE nexus method application in the continent highlights the demand for actionable information to enhance the production of water, energy, food, and ecosystems (Muhirwa et al., 2023).

Research on the FEWE nexus in African countries has recently discovered an extensive diversity of exciting perceptions, particularly in the FEWE resource network, consumption, and organisation (Botai et al., 2021). However, only a few articles indicate utilising the nexus method in resource production, while others focus more on synthesising each resource than their mixture (Hirwa et al., 2022). Such a definite study for bottom-up methods in the FEWE nexus has been demonstrated to be highly required. For instance, agricultural investors sustain economic growth and more user resources in Africa. Yet, these small-scale research studies may restrict the accessibility of information and knowledge regarding FEWE securities for policymakers at the peak decision levels. This requires knowledge of emerging trends in resource administration, particularly for the SDGs that need important national efforts.

3.2 POPULATION AND ECONOMIC GROWTH IN AFRICA

Population growth and economic enterprises are leading promoters of requirements in Africa. Fast demographic variation will undoubtedly remain a motivator for FEWE energy demand services throughout Africa in the coming decades. In 2020, the population of Africa accounted for over 1.3 billion people, nearly a fifth of the global population (Commission, 2021). This account was only 13% recorded in 2000, recounting the much quicker population rise on the continent than the rest of the globe. Africa's population has increased by an average of 2.5% in the past three years, quicker than the rest of the continent and more than double the worldwide rate. Only four countries, including Egypt, Nigeria, the Democratic Republic of the Congo (DRC), and Ethiopia, share approximately 40% of Africa's population. Nigeria remains the most inhabited country, with a population of about 210 million.

It was recorded that Africa had the youngest population in the world in 2020, with an average age of about 20 years, and more than 40% of the population are below the age of 15 years, which is significantly more than the global median of about 30%. The population in Africa is likely to grow by roughly 350 million, above the present inhabitants of the United States, to 1.7 billion by 2030. Nearly one in two individuals in the world between 2020 and 2030 will be African. Out of the 15 countries that will experience a significant population increase at that time, eight are in Africa, as presented in Figure 3.2. It is projected that in the next few years, the population in Africa will surpass that of both China and India. The African population is predicted to increase by the end of the century. The population of people living in rural areas was about 750 million in 2020, over 55% of Africa's population, one of the smallest developed continents in the world. Generally, the rural population has limited access to infrastructural facilities, such as energy, and fewer economic opportunities.

Nowadays, in African settings, the majority of the population remains in rural areas. However, the population in the city has been growing at an unparalleled rate, twice as fast as in the remaining part of the world. The urban population is projected to reach 35 million people by 2030, thus sharing the growth in population, including city migration. In view of this, it is expected that there will be an extra 40% accommodation for teeming migrants in African urban areas in comparison to the present situation. However, over one million people live in 67 African urban areas, and the three mega-urban regions have over 10 million people in Lagos, Kinshasa, and Cairo. Cairo in Egypt is predicted to be the most populated urban area, seconded by Kinshasa in the DRC and then Lagos in Nigeria. At the same time, Dar es Salaam and Luanda tend to be megacities by 2030, as shown in Figure 3.2.

The fast and rapid urbanisation has prompted the rate at which the FEWE nexus is demanded in Africa. It is observed that the establishment will progressively be

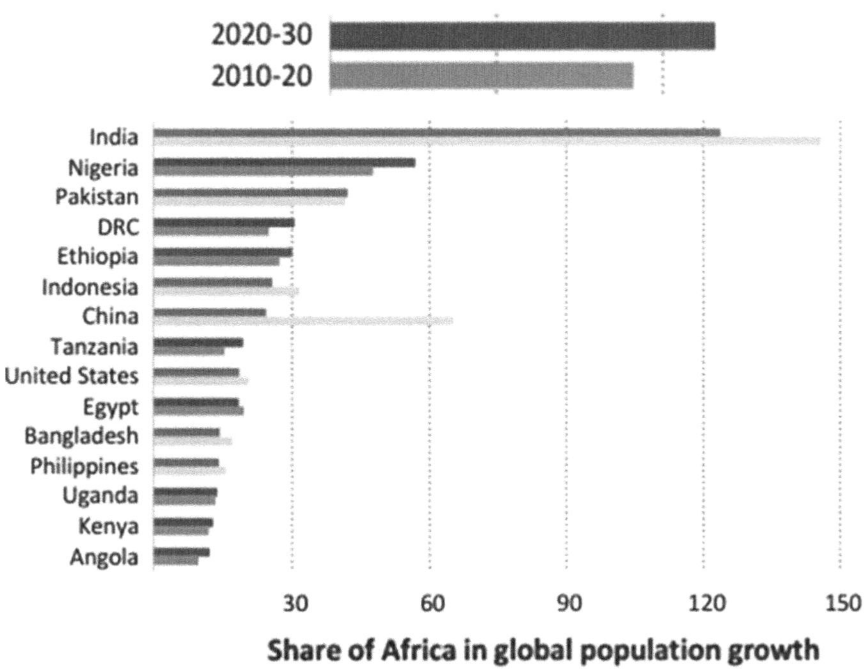

FIGURE 3.2 Population growth in the top 15 countries of Africa and its contribution to the global community, 2010–2030.

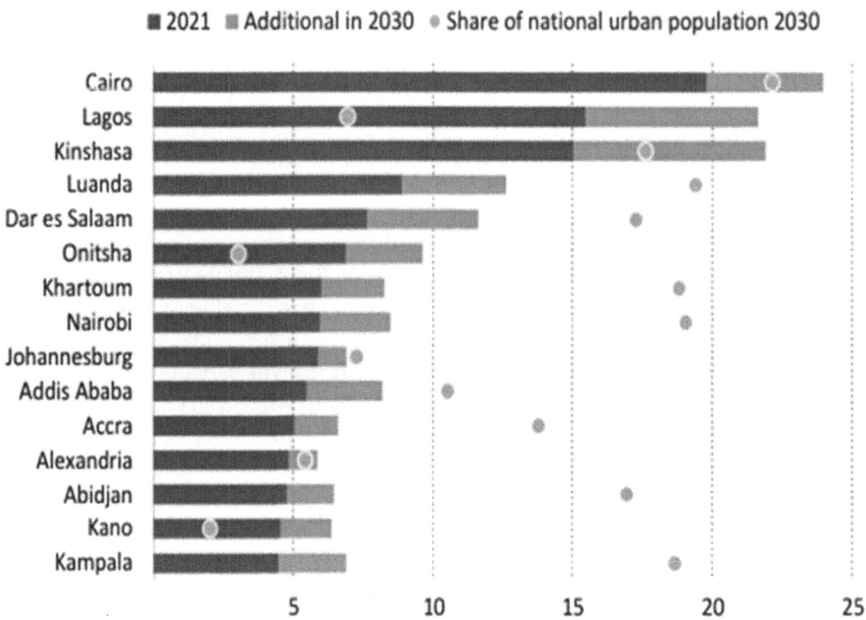

FIGURE 3.3 Population in the 15 major cities in the continent of Africa, 2021–2030.

significant in determining the entire FEWE inclinations on the African economy and the world in general as presented in Figure 3.3.

The short-term economic predictions for Africa are highly undefined. Many countries on the continent have deteriorated their economy due to the COVID-19 pandemic, which has ravaged their financial status worldwide. The economic recovery from the outbreak has been more robust than earlier projected in some of their countries, particularly in Southern Africa. The prices of commodities such as fuel, food, and fertiliser promote the earnings of countries where they are produced. Unfortunately, the financial prosperity of those countries has been adversely affected, especially at the beginning of the COVID-19 outbreak. Consequently, this weakens the economic returns, causing a rise in the African continent's health sector, debt, and joblessness.

Notwithstanding these sharp near-term unrests, longer-term opportunities in the economy of the continent are still dependent; as previously estimated, the GDP target by 2030 is expected to be approximately 50% greater than in 2020, hence getting up to around USD 10 trillion. GDP increases an average of 4% more than 2020–2030, a level similar to that for developing markets and emerging economies and 0.6% more than the world average. Expected increase rates differ significantly around Africa, such as in sub-Saharan Africa, except South Africa, which is the most rapid, with its contribution to overall GDP in Africa increasing from 55% in 2020 to 58% in 2030; the contribution of South Africa begins to decrease, from 11% in 2020 to 9% in 2030. Also, in North Africa, the contribution is around one-third. The segmental shares of GDP differ among countries, but the whole design of the economy in Africa is established to continue essentially unchanging for more than the next decade.

Energy utilisation in Africa accounts for about 6% of the world's energy use, irrespective of the fact that it occupies 16% of the global population. This condition is severe in sub-Saharan Africa, where over two-thirds of people have poor modern energy resources and about 60 million people cannot access electricity. However, the continent demonstrates the quickest population growth globally. For example, the population in sub-Saharan Africa from 2000 to 2012 grew by 270 million, and an increase in energy usage of 45% was observed (IRENA, 2017). The entire energy supply mix shows robust changes concerning the country or region. North Africa mainly relies on natural gas, about 50%, and oil, accounting for 47%, whereas South Africa depends mostly on coal, accounting for 70%. The scenario in other parts of sub-Saharan Africa is entirely dissimilar, with biomass and waste recording about 76% in the energy mix and minor shares from oil of about 16%, natural gas recording 4% and the rest of renewables recording 2.5% (Birol et al., 2014). Fuel types by total primary energy supply and the installed capacity of power plants in Africa are presented in Figure 3.4.

In 2021, about 600 million people, accounting for 43% of people in Africa, had no access to electricity, whereas 590 million came from sub-Saharan Africa as shown in Figure 3.5.

The potential energy forecasts in Africa correspond with the growing drifts experienced some years ago. The region will likely face the quickest rise in energy utilisation globally, motivated by population growth, GDP growth, and urbanisation ('Africa Energy Outlook 2022', 2022). In 2040, the population in sub-Saharan Africa is projected to virtually double and overall primary energy utilisation is likely to grow by approximately 80%. Moreover, the power sector is expected to quadruple its production capacity, the electricity demand will triple, and more than 950 million people will have access to electricity (Sanchez et al., 2020). This projected growth in energy generation would suggest an additional burden on African water resources and might increase the rivalry with the agricultural sector.

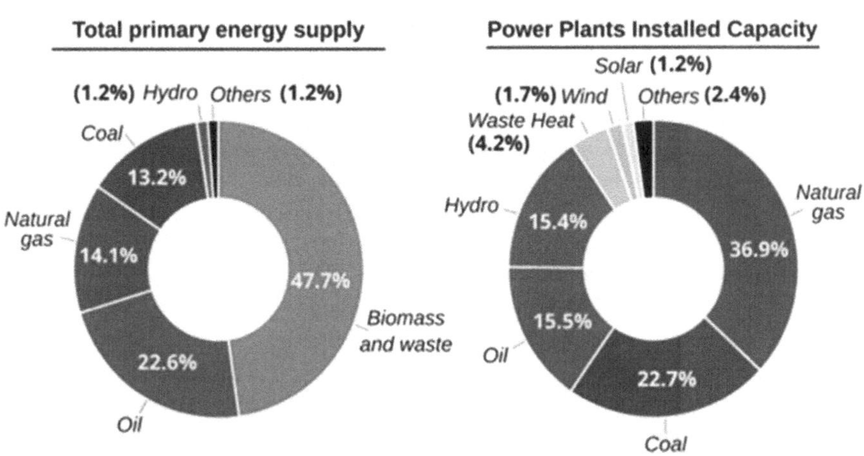

FIGURE 3.4 Fuel types by total primary energy supply and the installed capacity of power plants in Africa in 2016 (Sanchez et al., 2020).

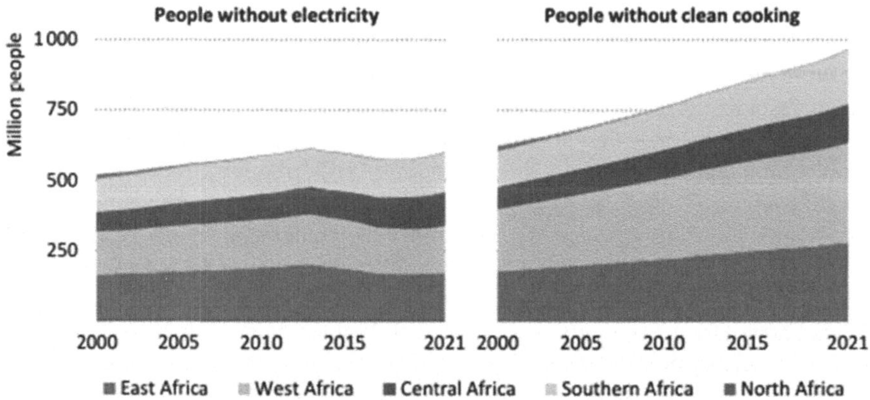

FIGURE 3.5 Population without access to modern energy services in Africa ('Africa Energy Outlook 2022', 2022).

Presently, about 86% of the water extraction in Africa is in demand for agricultural purposes, 10% for municipal uses, and about 4% for industry and energy ('Africa Energy Outlook 2022', 2022). Lack of water affects several places in the continent. Two-thirds of Africa is arid or semi-arid, and over 300 million people living in sub-Saharan countries are influenced by harsh water scarcity (Niasse and Varis, 2021). Africa occupies 20% of the global land (Muraina et al., 2023), but it contributes only 9% of renewable water resources. Besides, these water resources are unequally distributed among regions or countries. For example, North Africa, occupying 15% of the African land, is the most dehydrated area with a typical precipitation of about 96 mm/annum and a 1% contribution to the African renewable water resources.

In contrast, occupying just 18% of the land, Central Africa shares about 48% of the renewable water resources in Africa (Sanchez et al., 2020). These huge inherent variations could be caused by variations in precipitation and climate patterns because of global warming. By 2030, precipitation will likely intensify by 50% in some places and reduce to about 33% in others (Sanchez et al., 2020). High stern and lengthier floods, droughts, and lower river drifts in the dry seasons could be some of the effects of the previously huge drylands on the continent.

On the other hand, this chapter proposes a context for enhancing FEWE securities in Africa, which are characterised by unequal resource circulation and competition. The analysis examines the benefits of solidifying alliances among FEWE transboundary nexus systems. This partnership is enabled through common food trade, energy networks, and water basins, contributing a feasible remedy to regional competitive challenges. The key target is to evaluate the possibility of creating coalition approaches among transboundary nexus systems within Africa. The chapter stresses the advantages of engaging the FEWE nexus system tactics to distribute resources sustainably and proficiently between resource-scarce and resource-rich African countries. To demonstrate this, a case study is presented, addressing different likely inter-sectoral and intra-sectoral connections. This study represents a novel contribution to ensuring

FEWE securities in Africa. It presents a distinguishing method aimed at attaining regional-level stability and resource security. Ultimately, security and resource resilience could be achieved and increased at the national level, contributing a new viewpoint on solving the challenges related to FEWE in Africa.

3.3 WATER-ENERGY-FOOD RESOURCES IN AFRICA

Hydropower remains among the energy sources highly influenced by droughts, and several million people within the continent rely mostly on it, particularly in the sub-Saharan region. In 2016, the total installed hydropower capacity in Eastern Africa was about 54%, while 58% was recorded in Central Africa. Western Africa accounted for 30% (excluding Nigeria, which depends on natural gas), including 14 other nations recording more than 50% hydropower and eight countries above 70% (Aliyu, Modu and Tan, 2018). These countries that mostly rely on hydropower have experienced significant electricity shortages because of the poor water supply brought about by harsh droughts (da Silva, de Marchi Neto and Seifert, 2016). Conversely, floods can influence hydropower production, which may reduce dam capacity because of siltation and the demand to grow the resources' flood security boundary (Bejarano, Jansson and Nilsson, 2018). The water loss attributed to evaporation is an essential factor in the overall water intake from hydropower plants. However, the intake volume can several times exceed water consumption in nuclear plants or fossil fuels for each unit of energy produced (Macknick et al., 2012). The losses by seepage are illustrated as an extra source of water intake.

Nevertheless, water loss due to seepage usually occurs at the water reservoir, whereas evaporation represents an actual loss of water. Water loss through evaporation and seepage has been examined internationally and locally for the past two decades. Adopting water challenges, as it is usually performed for other fuel categories, is an unsuitable procedure in terms of hydropower because of the difficulty of the parameters that influence the connection between evaporation and energy generation of a hydropower plant. The challenges stated in the literature are categorised by significant inconsistencies (Mekonnen and Hoekstra, 2012) in the assessment of water shortage, resulting in overestimation or underestimation of overall water deficit. Previous studies on water utilisation for hydropower have been conducted at the international level (Hogeboom, Knook and Hoekstra, 2018) using basin data from international databases (Lehner et al., 2011) or focused on case studies for which highly accurate data of the basin parts may be needed (Camposeco-Negrete, 2013). International databases offer a wide range of data for many basins around the globe.

Nevertheless, there is limited information about when these places were assessed, and little is known about how they have developed over time Procurement of reservoir water surface data for sustaining spatial and temporal purposes frequently signifies an essential issue at the international level when the number of basins is significant. The volumes and surfaces of hydropower reservoirs often differ over time because of rules (storage, retaining, and discharging water), sectional-exact climate (evaporation, droughts, heatwaves, wind speed, and precipitation), and the intensity and number of shared users (Busker et al., 2019). A few problems indicated in Africa regarding the energy-water nexus are important. Moreover, the poor administration of existing water resources is causing water disasters in Africa. Fluctuations in climatic patterns are likewise anticipated to affect crop productivity, which, if joined with population growth, will result in extra pressure on water resources that must be devoted to growing agricultural output. In these circumstances, potential water requirements from the increasing African energy segment may play a crucial part when combined with variations in water accessibility and potential growing needs from agriculture. Thus, an appropriate study of water utilisation of the African energy sector is significant for the operative potential preparation and organisation of energy and water resources in Africa.

3.4 PAST AND PRESENT STATES OF THE FEWE NEXUS IN THE CONTINENT OF AFRICA

The African population is approximately 1.7 billion, equivalent to 16% of the global population. A few decades ago, food and energy demands rose tremendously in line with the increase in population and economic growth. The production of crops increased by over 35% from 2006 to 2016, whereas electricity demand rose by over 20% simultaneously. The present production in agriculture is mostly based on rainfall. Flood recession, irrigation, and crop cultivation are mainly done in dry places along the major rivers in Senegal, Niger, Sudan, Mali, Zimbabwe, Egypt, Namibia, and Nigeria. Only a small portion (roughly 5%) of the poorly accessible water resources are presently reserved, mainly in the Northern part of the continent and primarily for agricultural purpose. A significant advancement in modern irrigation services is anticipated in the coming years (Gabr, 2022). Electricity in major parts of Africa is generated from a grid prepared in five power pools (Eastern, West, North, Central, and Southern African Power Pools, respectively), with a target to organise power network development and operation within their member nations as presented in Figure 3.6 by developing inter-relationship curves and a regular application of market-centred combined methods (De Felice et al., 2019; Busch, De Felice and González, 2020; Pavičević et al., 2020; IRENA, 2017).

In Africa, hydropower is recognised as a key renewable energy generation source and the second representative source after fossil fuel-powered generation from thermal plants (Farinosi et al., 2017; Sanchez et al., 2020). Nevertheless, the continent remains the most underutilised hydropower resource in the world, with estimates indicating that only 11% (37 GW) is currently utilised. Significant investments aimed at expanding hydropower capacity are expected in the near future, and they will be converted into more than 200 modern installations (Zar et al., 2015). The major aspects of these modern facilities are organised to be executed in the major wettest places and the water basins in Africa. Besides the hydropower expansion, various planned water transfer schemes are being considered in the major basins of the continent. A number of basins are in a developed region, such as the "Lesotho Highlands Water Project" between South Africa and Lesotho or the "Mzimvubu" and the "Mokolo and Crocodile River Water Augmentation" schemes in South Africa. Further projects are found in a smaller, more limited stage of implementation such as the "New Nile" and the "Transaqua" schemes, with the design to separate water from Lake Chad and towards the Congo basin to the Nile, respectively (Shumilova et al., 2018). Further water

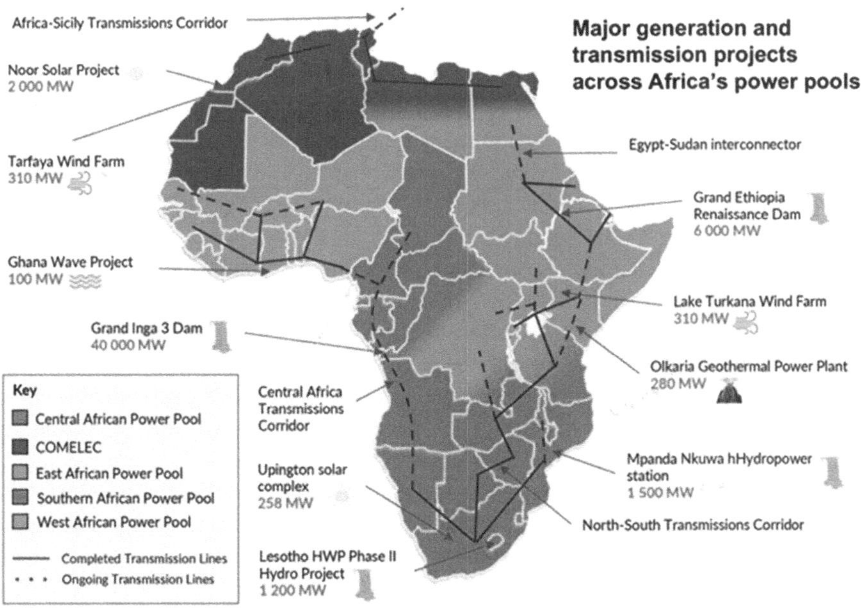

FIGURE 3.6 African power pools. (Adapted from United Nations Environment Programme, 2017.)

transmission projects were planned for the Zambezi basin between Botswana, Zimbabwe, Namibia, and Zambia.

On the other hand, the deficiency of accurate and reliable databases is a significant obstacle to the technical application of the FEWE nexus (Orimoloye, 2022). Presently, no devoted organisation is in control of collecting FEWE nexus information, creating a consistent database and offering links to water-connected enterprises, events, and papers. This condition requires improved assistance from diverse regions of the continent. Many countries experience challenges with low levels of data accessibility and accuracy, making it hard to conduct wide-range local and international FEWE nexus analyses that necessitate thorough socio-economic and climate information. Furthermore, certain countries may be uncertain to share certain types of essential data because of national strategic policies and security considerations (Markantonis et al., 2019). A lifecycle perception should be employed when assessing various strategic possibilities to prevent this condition. Industry, perhaps, practices lifecycle procedures and measures to resolve water challenges. The impacts of water security extend beyond manufacturing sites and affect the entire supply chain. mainly in the case of agricultural products, which often involve substantial quantities of water and energy for management and harvesting (Apeh and Nwulu, 2025).

3.4.1 Water-Food Nexus

Population growth in Africa has been predicted to increase due to increased demand for agricultural outputs such as crops. Following the current viewpoint (OECD, 2018), the population in Africa is likely to increase by over 300 million inhabitants in 2030, furthering the call for agricultural outputs and growing the generation by nearly 21% with regards to the net worth added (around 16% characterised by crop production and 5% by fish and livestock farming). Tubers, roots, cotton, and grains likely characterise most crop yields. Prolonged agricultural production is anticipated to be mostly the output of strengthening, cropping, and higher production competence, but it will also raise the call for land in agriculture, where places of harvest are likely to increase by about 4 Mha. The demand for animal protein, such as poultry, ovine meat, and bovine meat, is predicted to rise gradually in the coming years. The resulting advancement in production is mostly permitted by a transformation of the agricultural supply chain and systems, but it will also bring a projected 18% rise in GHG discharge from the agricultural sector.

The rise in population has become a major issue influencing the growth of sectors, mostly in agriculture. Still, per capita consumption is likely to rise by 75 Kcal/day, attaining approximately 2,510 Kcal/day. Notwithstanding this, it is worth the smallest global unity, and it increases per capita consumption of Africa to an average of around 80% globally (OECD, 2018). Moreover, on average, each country in the continent had about 0.9 mm of agroforestry-related precipitation in 2000, which increased to approximately 48 mm per year by 2020 (Muhirwa et al., 2023). The countries that receive the highest rainfall from forests are South Africa, Angola, and Ethiopia.

The food production market in Africa is anticipated to mostly benefit from the execution of the African Continental Free Trade Agreement contracted in 2019, guaranteeing a decrease in operational charges by over 90%. The key limitations to agricultural expansion are inadequate means to technology, shortage of facilities, the disintegration of appropriate arable regions, insufficient water supplies, huge weather uncertainties, and restrictions to means of irrigation systems where arable land is mainly accessible (OECD, 2018). The creativity to enhance the viability of agriculture has been previously predicted in different African nations. For instance, in Kenya, the study of the effect of agricultural rules (emphasising fertilisers) exhibited a widespread reduction in food prices. It encouraged access to food impact for every family in the state, accompanied by a rise in food intake in every modelled situation.

Nevertheless, this rise varied across families along with their revenue, nutritional energy intake, nutritional variety, and children's dietary position (for instance, in the poorest families, the highest growth in macronutrient consumption originated from proteins, whereas for average-income families, it originated from carbohydrates) (Ramos et al., 2020). Furthermore, the agricultural commercialisation cluster (ACC) has been presented in Ethiopia by the government as a device to advance agricultural output and production in more possible areas for a few important commodities (wheat, teff, maize, and barley) (Louhichi et al., 2020).

3.4.2 Water-Energy Nexus

In most African energy systems, hydropower serves as the main renewable energy resource. A study by the World Bank highlights that water and electricity demands are expected to increase meaningfully by 2050 in the continent by 700% and 500%, respectively, compared to 2012 (Rodriguez et al., 2013). Other prominent features of the projects are their low electrification charges (Meyer, Apeh and Overen, 2020), high oil shares in the power production network, minor sizes, and the shortage of substantial gas and power relationships. Power systems are limited by the temperature and accessibility of water resources, as thermal power plants require water for cooling and the generation of hydropower depends on water to function.

Thus, weather uncertainty robustly affects energy networks in Africa (Apeh, Overen and Meyer, 2021), influencing their energy network, working prices, carbon dioxide (CO_2) discharges, and water intake for energy production. For instance, in the Eastern, Central, and Northern African Power Pools, changes between wet and dry years could differ in the energy sector emanating from hydro units equal to 5.2%, initiating fluctuations in the working prices of about 1.4 billion € (or 3.8 €/ MWh) and encourage rotations in CO_2 discharges equal to 15 million tons per year (Pavičević et al., 2021). In addition, power systems can affect the amount and quality of water resources. For example, the losses in water because of production in hydropower in Africa cannot be overlooked, which is approximately 42 billion cubic metres (BCM) in 2016 (corresponding to the total of the other fuel types combined, the projected volume was 1.2 BCM) as shown in Figure 3.7.

Compared to the nations with the greatest water damage because of hydropower reservoir evaporation, Ghana recorded over 2,322 BCM of water loss from evaporation and sewage. This loss is greater than the combination of energy being produced

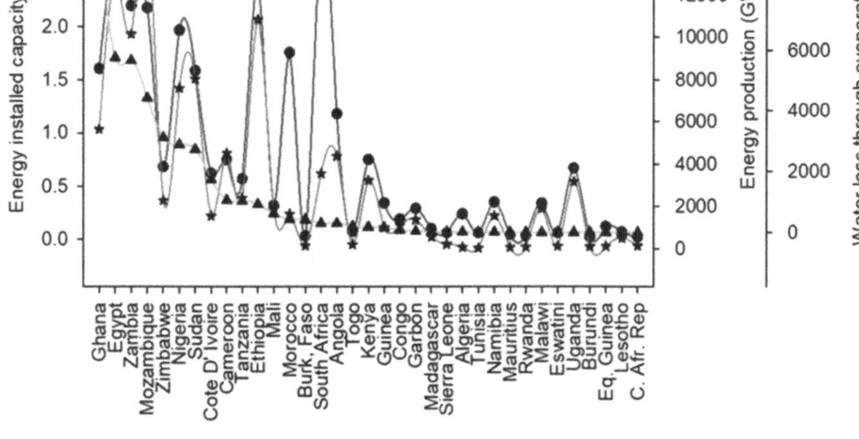

FIGURE 3.7 Energy installed, production, and water loss (evaporation and seepage) of 35 countries with hydropower.

from other fuel sources. Similarly, at the Aswan Dam in Egypt, hydropower consumption is secondary, with only 2.7 GW installed capacity compared to the 36 GW for the remaining installed power plants on the continent. Nevertheless, the water loss associated with hydropower generation is 33 times greater than the combined water consumption of all other fuel sources. Meanwhile, applying non-hydro renewable energies as a substitute for fossil fuels can meaningfully decrease water consumption while encountering the rising energy demands of the continent (Sanchez et al., 2020). Furthermore, grid networks could significantly assist in reducing water extractions and energy generation intake in connection with the current system designs and the regular electricity charge (Busch, n.d.). The energy combination in Africa will likely experience substantial deviations in subsequent decades, as presented in Figure 3.8. Fossil fuels such as carbon, oil, and gas will remain significant energy-generating inputs, with oil and coal anticipated to shrink considerably.

Moreover, important changes at the district level are projected in terms of other sources of renewable energy. While Southern, Northern, and Eastern Africa could acquire this modern energy from wind power, geothermal sources in East Africa, solar photovoltaics (PV) in both the Southern and Northern districts, concentrating solar power (CSP) will be significantly effective in North Africa (IRENA, 2017). In this multifaceted context, geothermal energy, which presently accounts for a marginal portion, will nevertheless encounter a great increase in energy production in 2040. Among the major benefits of this energy resource is its little effect on the environment and GHG discharges concerning energy produced through fossil fuels, a

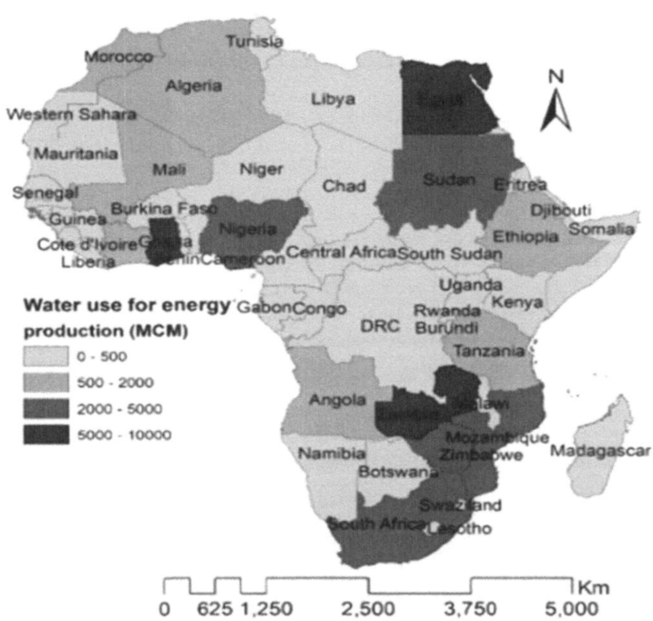

FIGURE 3.8 Average energy-water nexus production in Africa from 2000 to 2020 (Muhirwa et al., 2023).

fairly continuous production yield independent of climatic conditions (which makes it mostly appropriate for base load electric production), a viable levelised cost of electricity (LCOE) production, and less susceptibility to the uncertainty of the international Oil and Gas (O&G) commercialisation.

The East African Rift System (EARS) generates more constructive situations for geothermal energy at cheap and practically drillable pits (below 4,000 m), with a universal possibility projected at 15,000–20,000 MW. Nevertheless, presently, a small share of the geothermal system has been achieved in Kenya due to the strong and clear governmental policies, the absence of regional practical and administrative skills, the inaccessibility of several geothermal places compared to O&G areas (where many of the drilling workers and facility benefactors are built), the insufficient funding during the initial phases of the schemes, the challenges from other energy sources, and the problem of compensable charge for the produced electric power in still less recognised in national electric markets (Atheru, 2018). Notwithstanding the seemingly restricted significance of geothermal energy, it is good to record that a limited number of nations uphold the majority of the possible, and a few minor countries could produce most of their electrical energy demands (e.g., Djibouti, Comoros).

3.4.3 Food-Energy Nexus

Globally, food production records about 30% of energy consumption (Wieben, 2016). Energy applications are essential in every stage of the food production chain, from the farm to treatment, processing, farm storage, circulation, and arrangement for use (Kanda et al., 2023). Poor accessibility to energy sources in African countries remains a major reason for the losses of farm products after harvest in the agricultural sector and, hence, a significant cause of lingering food uncertainty and starvation (Lynd et al., 2015). A typical illustration of a straight relationship between food and energy production is biomass energy (Tashtoush, Al-Zubari and Shah, 2019). Hence, biomass energy dependency by homes and precisely in Africa Siren makes people vulnerable to weather fluctuations and changes. Similarly, feedstocks utilised few irrigations, while most were rainfed for biofuel production. Thus, the effects of changes in weather on energy generation from agricultural biomass can never be neglected. Bioenergy generation involves land and is therefore inseparably connected with social growth, environmental quality, and agriculture, which can result in unwanted consequences if the connections are poorly and inaccurately managed (Lynd et al., 2015).

In Ethiopia, for instance, biofuel generation from sugarcane molasses, castor beans, and *Jatropha* has increased social safety by bringing out more job opportunities, even though the danger of food insecurity has risen because of land resource competition (Kanda et al., 2023). Using land for crop production and biofuel has reduced the land for food crops. Also, Ethiopia has assigned 23 million ha and 700,000 ha to international firms for *Jatropha* for biofuel generation and sugarcane cultivation, respectively (Dobaa, Abbas and Metwaly, 2014). Using animal droppings for bioenergy directly competes with their use as dung in agriculture, resulting in reduced agricultural crop yields. The challenge of nourishing a growing population requires increasing food production from the prevailing agricultural land or expanding the areas under crop cultivation. As many places in Africa are either arid or semi-arid, rainfed agriculture is exposed to weather variation and uncertainty, so irrigation is needed. It needs energy to drive and circulate water; hence, increasing irrigation areas increases energy needs. For instance, the demand for large-scale irrigation in Ethiopia's electricity projects is growing and is expected to reach around 13,000 GWh in 2030 (Yalew, 2022). As described earlier, hydropower is exposed to weather uncertainty and variation, affecting irrigated agriculture. Implementing weather-smart energy applications and administrative exercises can support weather variation transformation and reduction, as stated in the framework of Ethiopia (Gebreyohannes and Zenebe, 2019). These performances can be used in different parts of African countries.

3.5 CLIMATIC DESCRIPTION OF THE AFRICAN CONTINENT

Africa is characterised by mostly uncommon precipitations irregularly distributed over time and space (Aude et al., 2017). Places characterised by moderately high and plentiful rainfall are on the Western African coast, Ethiopia, Lake Victoria, the Gulf of Guinea, and the Equatorial Congo. Apart from most arid zones, the precipitation patterns are categorised by a stout periodicity. Inter- and intra-yearly weather fluctuations are high and bring a considerable variation regarding the organisation of natural resources in the major water-concentrated anthropological events, such as agriculture (Nicholson, Funk and Fink, 2018). The Sahelian area has an attractive, unvarying precipitation design. After the 1960s, the precipitation seriously weakened in terms of the moderately high rich first portion of the 20th century, declining slightly towards growth after the 1980s (Nicholson, Funk and Fink, 2018). The oceanic fluctuations deeply influence precipitation differences in sea surface temperature.

After the 1930s, considering the rainfall accounts, a decadal uncertainty is evident in the amount of rainfall; the Northern and Western areas of the continent encountered a substantial descending drift in rainfall in the previous decades, with situations drastically fluctuating after 1968 (Nicholson, Funk and Fink, 2018). Following the 1977/1978 La Nina-driven shift, the descending drifts stretched to a large portion of the continent in the Pacific equatorial zone. The Africans are divided based on the agroecological zones. The gradual rise in arid environments is seriously influenced by the sub-Saharan African population's already dangerous food security situation. Considering the expected precipitation drifts in prospective decades, the condition appears unchanged for high constructive conditions (Dosio et al., 2019). The rainfall commands are not expected to differ significantly in the northern part of the continent. In addition, systems do not accept the likely variations in mean precipitation over Western Africa: some procedures expected a little wetter, and others expected drier situations. The majority of them, however, indicate a decrease in precipitation occurrence and an increase in the duration of dry spells across the area. Eastern Africa is anticipated to undergo a slight increase in the mean yearly rainfall but with a mixture of more considerable intra-annual inconsistency, with high precipitation in the wet season and growing dry spells in the dry season. Central and, specifically,

Southern Africa are likely to experience a prominent decrease in precipitation drifts, both through yearly average and wet spells (Dosio et al., 2019). Heatwaves, drought, and dry spell situations are predicted to be more regular and stronger in the prospective decades (Dosio, 2017; Dosio et al., 2018; Spinoni et al., 2020), whereas more regular floods are predicted to hit the wetter shares of the continent (Alfieri et al., n.d.). The joint existence of the foremost hydro-meteorological hazards is likely to increase mostly, particularly in the areas where population variation is predicted to be more substantial; hence, the general danger is expected to increase (Migali et al., 2018; Sanchez et al., 2020).

3.6 PROSPECTIVE CHALLENGES TO SOCIO-ECONOMIC FACTORS REGULATING WATER USAGE IN AFRICA

3.6.1 Implications of Climate Excesses under Various Demographic Variations, Urbanisation, and Global Warming Conditions

Regarding the latest socio-economic predictions in Africa, it is likely that the continent will experience significant growth in its population (which is projected to go beyond 1.7 billion inhabitants in the prospective decades until 2050. In addition, Africa has some of the least developed populations in the world (77% are presently below the age of 35, and two out of every five children globally will come from Africa by 2050), and the number of people is projected to be three times by 2050 (Hajjar, 2020). The diverse connections between development, socio-economic growth, and migration possibilities are expected to significantly affect demographic changes in literacy, distribution of economic resources, fertility, employment, mortality, and age distribution.

Still and though the shared socio-economic pathway (SSP) set-ups (Riahi et al., 2017) define these possibilities of socio-economic prospects (bearing in mind the operation of alleviation laws or their absence), their efficacy could be reduced since they do not include the chance of development interruptions (Buhaug and Vestby, 2019). This could be mainly significant in the framework established by the existing COVID-19 epidemic. In view of the shared socio-economic pathway 3 (SSP3) set-up, the one categorised by sluggish financial growth, persistent disparities, and material-intensive consumption, the population growth is projected to be considered by a slower step in commercial and more in emerging nations and a very strident one in the emerging world. Furthermore, the situation is described by scarce environmental anxieties, bringing about robust environmental dilapidation in some areas. Because of this situation, the population in Africa is predicted to grow quickly, particularly in the Central-Eastern and Western parts of the continent.

Climate change is anticipated to influence the African continent, especially the previously tense water resources. The continent's socio-economic development mainly influences climatic growth. Globally, most changes in the human population and a great portion of the economic growth will likely influence the African continent in the subsequent decades. This represents an initial assessment of how the affected population may experience and respond to potential changes linked to Africa's most significant hydro-meteorological events under three different global warming level (GWL) scenarios. Several studies have been conducted in the field (Alfieri et al., n.d.; Dosio et al., 2018; Naumann et al., 2018). The work examined the experience of present and upcoming populations in Africa to a growing proportion and existence of inflamed hydro-meteorological risks, for example, drought, flood, and heatwave under the Paris objectives, respectively, suggesting 1.5 and 2°C progress in global mean surface temperature (GMST) over pre-manufacturing phases (Frieler et al., 2017; Hoegh-Guldberg et al., 2019).

Similarly, to examine the possible charges of the failure to fulfil those objectives, deliberations were made to present a further action that designates a GWL of 3°C, more than the Paris objectives (Raftery et al., 2017). The estimations of the extremes in weather were integrated with approximations of the increase in population from SSP3 (Riahi et al., 2017). The growth concerning risks were examined as shown in Figure 3.8, and jointly, it was used to expand and recognise how the relationships existing in different risks, anticipated to support one another, could endanger the coming generations of Africa. The study regarding many risky mixtures was mostly recognised in the work, while disseminating the percentage of the population prone to risks and the work of the accurate variations of the three mixtures. In West Africa, for example, if a normalised measure of 0 is considered as the least multi-risk proportion while the maximum is taken as 300, then, in the zero states, 100% of the people (about 300 million people) are likely to be the average of individuals prone to a multi-risk proportion, which is above 25 (interquartile range 0–40), while the highest exposure 10% (about 30 million people) may experience an amount higher than 100 (75–135). In an instance of 3°C GWL, 100% of the individuals (assessed to rise from ~300 to ~790 million people) may be liable to a multi-risk average proportion above 50 (15–70), while the highest disposed to 10% (~79 million individuals) would control a proportion of around 200 million (165–250). Mostly, the increase of the multi-risk susceptibility ranges from 2°C and 3°C, meaningfully above the ranges of 1.5°C and 2°C conditions.

The rise in the multi-risk susceptibility of exposed individuals in the Mediterranean Regions (1.5°C that is +10%, 2°C = +200%, and 3°C = +400% GWC in terms of the site condition), West Africa, and North-East (with a similar increase of about +200%, +300%, and +700%, respectively) outcomes arise from the alliance of differences in both the risks of incidence and susceptibility of individuals. In order to divide the GDP and the number of people prone to several stages of the three risks, the increase in the GWL is clear in all conditions on every continent of Africa. Averting global warming to the Paris Agreement temperature aims (1.5°C and 2°C, respectively) may significantly reduce the proportion of the people in Africa vulnerable to three prevalent hydro-meteorological dangers—floods, heatwaves, and droughts—against a 3°C warming situation. In reality, it is observed that maintaining a temperature rise of less than 2°C may decrease the proportion of people liable to the mixed excesses of hydro-meteorological activities by over 50% in Africa. Extra work to restrict it to 1.5°C may additionally decrease the susceptibility by around an extra 10 to 30%. The greatest susceptibility of the people in Africa to dangerous hydro-meteorological conditions could worsen the challenges connected to water control in Africa (Farinosi et al., 2018; Migali et al., 2018; Sanchez et al., 2020).

3.6.2 Climate and Socio-economic Factors in Serious Water Regulation

The Armed Conflict Location and Event Data Project (ACLED) gathers records of political conflicts and demonstrative actions around third-world countries. This political data included dates, involved parties, nature of violence, locations, casualties, and brief accounts. Remarkably, ACLED was not initially personalised for recording cases of violence associated with water resource consumption and management. A subset of the wide dataset, encompassing more than 162,000 occurrences, was formed to recognise such cases. This subset was created by engaging and establishing keywords linked to numerous facets of water organisation, such as water availability, pricing, utilisation, sanitation, agriculture, irrigation, hydro-meteorological hazards, and more. Specific care was taken to prevent measures where water was employed to quell protests, such as using water cannons or territorial disagreements involving marine regions.

Consequently, the water-connected measures removed from ACLED records offer insight into water management issues in certain African places. While the circulation of these measures shows obvious temporal variations, different spatial clusters can still be recognised. The key regions with sensitive incidences of water-related violence are outlined as high density (HD). It is vital to note that the nature of these issues is changing across countries. Although the most regularly stirring terms in events describing a definite country or region may change meaningfully, they usually spin around a few key themes, such as natural hazards, water supply, sanitation and hygiene, and water-intensive economic activities like mining and agricultural application of water resources. Moreover, the nature of water-connected violence incidents varies between rural and urban areas. Urban areas mainly contend with water supply and sanitation issues, pricing, and resource availability, while rural areas are affected mainly by challenges connected to agricultural water usage.

3.7 ZAMBEZI AND NIGER AS CASE STUDIES FOR TRANSBOUNDARY WATER CONTROL IN AFRICAN RIVER BASINS

3.7.1 Zambezi

3.7.1.1 Introduction

The fourth largest African basin is the Zambezi, after the Congo, Nile, and Niger River Basins. The region's drainage is approximately 1.4 million km, passing across the borders of eight Southern African countries. The features of the Zambezi basin zone are the fluctuating weather conditions from the arid and temperate environment in the south to the subtropical environment in the north. The average yearly precipitation varies from 1,200 mm/yr in the rainiest regions of the Shire and Kafue subsystems to below 700 in the driest zones in Zimbabwe. The area is divided among the following nations (proportion of the region and hydrological balance: neutral [N], water consumer [C], and water producer [P]): Zambia (41.7% [P]), Botswana (1.0% [C]), Malawi (8.0% [P]), Namibia (1.0% [N]), Mozambique (1.8% [P]), Zimbabwe (15.6% [P]), and Tanzania (2.0% [P]). The Zambezi River basin is home to approximately 40 million people, with around 32 million living in rural regions. The Zambezi Watercourse Commission (ZAMCOM), founded in 2011 by the eight riparian nations, directs the basin's harmonised transnational and transboundary strategic management (Sadc, Zamcom and Arendal, 2012).

3.7.1.2 Major Challenges

After the stormy era of decolonisation largely characterised by civil war, which was observed in Angola up to the past decade, the countries in the Zambezi River basin encountered an era of quick socio-economic advancement. The basin is mainly endowed with natural resources. Major commercial activities are characterised by fisheries, manufacturing, extraction, agriculture, and tourism. Even though the region's commercial activities are enhanced by mining essential resources such as copper, oil, diamonds, coal, and cobalt, most people are still malnourished, suffer from essential amenities, and have high infant and maternal death rates and short lifespans. The major commercial activities that boost the economy of the region are the service sectors, industry, and agriculture. Specifically, agriculture regulates approximately 17% of the entire GDP of the nation, donating the basin of the region (from 2% in Botswana to 29% in Mozambique), whereas the industry contributes about 32% (from 19% in Malawi to 57% in Angola), and the service part shares 51% (from 32% in Angola to 60% in Botswana) (Sall et al., 2020). The population of residents in the basin was approximately 40 million, with three-quarters of people located in rural areas, such that Zambia shares 50% and Malawi shares 85%, while the remaining 25% is in the 21 leading cities. In Botswana, about 30% of the population is poor, while the poverty rate in Tanzania is nearly 90%; the means to fresh water is pretty small, about 50% in Tanzania, Angola, Zambia, and Mozambique. The per capita water intake within the basin averages about 101 m^3 yr^{-1}, fluctuating from 27 in Angola to 203 in Zimbabwe.

The main water-penetrating area is agriculture, of which over 76% of the resource is meant (42% in Botswana, about 60% in Angola and Namibia, and more than 90% in Zimbabwe, Tanzania, and Mozambique). Small-scale rainfed agriculture is the major source of income in the main basin. The essential crops grown in the area are cassava, wheat, rice, corn, cotton, groundnuts, sugarcane, tobacco, and sorghum. The land mapped out for agriculture in the basin was measured at around 5.2 million hectares, with irrigation of just 0.25 million. The advancement of irrigation facilities was recognised as significant in the Zambezi region, and plans to irrigate an extra quantity of 0.51 million hectares were previously recognised. Cattle production and fuelwood collection remain the major challenges to the natural ecosystems of the basin, which is very significant from a preservation viewpoint. Fisheries are the main source of revenue for the houses residing in the wetlands and the delta along the lakes' and rivers' coasts. The prospect of hydropower in the basin was projected to be around 20 GW, of which just around a quarter is achieved mostly in the Mozambique, Kafue Gorge dams and Cahora Bassa, in Zambia and Kariba. Approximately 40 extra feasible connections (totalling around 13 GW of connected capacity) have been previously recognised in the basin. Many of these proposals are being formed within the Southern African Power Pool (SAPP) advancement programme.

The proposed expansion will increase the utilisation of SAPP's hydropower capacity in the Zambezi basin, between 70% and 85% in 2030 (Conway et al., 2017). Different water transmission plans regarding the North-South Carrier or the Zambezi Water Transfer Scheme in Botswana have been previously proposed for the basin. Nevertheless, the appropriate conservation and restoration of the present facility is a source of foremost worry: Kariba dam is presently in an unsafe condition because of bedrock erosion (which looms over the constancy of the wall foundations), and its possible failure would involve disastrous costs across the basin (IRENA, 2017). Given the overall water shortage in the basin, weather variation and differences are regarded as the major pertinent pressures in the near future, both for the current human actions and for prospective growth.

3.7.1.3 The FEWE Nexus Applications

The study of the FEW nexus is not static as the African Network of Centres of Excellence on Water Sciences and Technology (ACEWATER) phase 2 mission focuses on the Zambezi basin as the key target of the African Union. Among the organisation's duties was the hydrological study of the effects of weather variability and growth on the water resources in the basin, which was executed in partnership with Rhodes University in South Africa. The description of the architecture system is designated in the study (Hughes and Farinosi, 2020), whereas an additional evaluation provides a brief summary of the case studies(Hughes and Farinosi, 2020). Some of the fluctuations in the experimental data applied in the model establishment and the connections with model output variability (mostly in terms of the simulation of a definite runoff-production procedure) are described by (Hughes and Farinosi, 2021). The effects of climate change and water utilisation set-ups will likely vary across the Zambezi River reservoirs. Most affected places are possibly characterised by exposed water bodies, natural lakes, and artificial basins, which are very subtle to various influences of rising dryness. The fluctuations for the prospective simulation outcomes vastly rely on a certain weather algorithm.

However, the outcome proposes a considerable reduction in prospective water accessibility for the combined conditions. The variations in the conditions of water utilisation have comparatively slight effects compared to the difference in weather effects, with the exclusion of some of the semi-dry sub-basins of Zimbabwe, where there are fairly concentrated water applications concerning existing water. These situations are anticipated to have a serious effect on the hydropower system set-up in the basin, and so the future water-connected threat in the entire SAPP (primarily dependent on the Zambezi basin for hydropower generation) grows. Concretely, the result from Zambezi hydropower could be reduced by 10–20% because of drier situations, while close tenure system production prices could rise by 20–30% for countries that rely on hydropower (Spalding-Fecher, Joyce and Winkler, 2017). Studies by Arias et al. (2022) show that flow changes are expected to be greatest at Cahora Bassa (−30 to −49%), followed by Kariba (−12 to −51%), Kafue Gorge (−25 to −42%) and Itezhi-Tezhi (−15 to −35%). Differences in the heights of water and flows could cause slight undesirable impacts on average hydropower production and rise in inter-annual variations; the highest fluctuations are projected at Kariba, where yearly production is anticipated to reduce from 5,989 to 4,152 GWh/yr (−28.1 to −0.7%).

At Itezhi-Tezhi, mean yearly hydropower is anticipated to differ from 675 to 534–536 GWh/yr (–21%), while at Cahora Bassa, mean yearly production (18,017 GWh/yr) is anticipated to reduce by 2%.

However, there are still other sectors that demand water; for instance, agriculture is one of the sectors that requires the highest water usage. Its drainage is nearly 1.4% of the basin's yearly renewable water activities. Approximately 90% of the entire farm regions are in the focal floodplains and are farmed largely with rainfed agriculture. However, the irrigated places are evaluated in 150,000 and 260,000 hectares. They are anticipated to triple by 2025, increasing the burden on the yearly renewable water resources to approximately 4.5% of the total.

3.7.2 Niger

3.7.2.1 Introduction

The Niger basin is bounded by nine West African countries, including Cameroon (4.4%), Mali (30.3%), Benin (2.5%), Côte d'Ivoire (1.2%), Nigeria (28.3%), Guinea (4.6%), Chad (1.0%), Burkina Faso (3.9%), and Niger (23.8%) with about 2.1 million km^2. The headwater is located in the Fouta Djallon Massif in Guinea, while the terminal delta is found in the Niger River in Nigeria (the third longest river on the continent). Niger Basin moves for over 4,200 km and forms an essential passage for movement and commercial activities, a prospective source of battle and an opportunity for cross-border collaboration (Chaturvedi, 2013). In conjunction with its related aquifer, it is recognised as one of the highest key water resources in Africa, providing access to water for consumption, income through livestock herding, a wide range of irrigation and fisheries with hydropower production, and tolerating navigation.

Furthermore, one of the biggest Ramsar locations globally includes the Inland Niger Delta, a hotbed of biodiversity, and an essential area of the ecoregional system (Zwarts et al., 2006). In the same vein, the basin is likewise enormously susceptible because of its obvious hydroclimatic fluctuations, both longitudinally and sequentially, along with the multifaceted socio-economic framework (high rate of poverty, increasing population, food scarcity, and standards of the Human Development Index (HDI)) showing the lowest recorded value worldwide (Vaucelle, 2015). In view of this, in 1980, the Niger Basin Authority (NBA) was instituted by the nine riparian countries, with targets to organise national water resource administrative efforts using a joint growth strategy for the basin to expand the living standards and affluence of people. Presently, the basin has an admiring prospect for advancement, as irrigated lands are expected to take only 9% for agriculture, while the installed capacity for the prospective hydropower is only expected to take 5% of the land.

3.7.2.2 Major Challenges

Water is a key element that contributes to the growth of agricultural products and sustainability in manufacturing organisations, such as power supply. Although, within the Niger basin, water reserves and demands are unequally circulated, riverine areas can be classified into water consumers, contributors, consumers and contributors, and lowest consumers and producers (Medinilla, 2017). Consequently, while the area is endowed with water reserves, it experiences an enduring water shortage and its entire renewable energy prospect is underutilised. In this scenario, the distribution of existing scarce water reserves for sustainable water application has been subject to growing interest, producing more water consumption and demand. Many available basins (more than 450 in the Nigerian aspect of the reservoir) are dedicated to flood regulation, residential use, irrigation, livestock farming and fishing, whereas only a small number were initially intended for hydropower production. More water reserve expansions should be cautiously deliberated in the Hadejia-Jama'are reservoir, which is found in northeastern Nigeria. Being one of the area's greatest and most prolific basins, the complete execution of the designed set-up, if not properly managed, could adversely affect the reservoir's floodplain swamps and the agricultural outputs of the regions found downstream. Products from traditional irrigation would not substitute for the decrease in farm output; meanwhile, the water reserves of the reservoirs may be unable to sustain the predicted scope of the systems.

In terms of the basin established in Northern Nigeria, weather fluctuations and their effects on the surroundings, such as food and water, are currently vast. The growing temperature and decline in rainfall have caused unvarying desertification and drought, disturbed many cultivable land places and decreased crop yield. This has encouraged huge relocations and migrations of people to places with little desertification effect. Moreover, due to climate change and uncertainty, the reservoir encounters several challenges such as land degradation, disaster vulnerability, water resources and related ecosystems, incompetence and inadequate performance of agriculture for both irrigated and rainfed, challenging demands between water consumers and sectors, and poor investment in water facilities. Generally, poor communal services, unemployment, poor governance and institutional failure, increase in population and urbanisation, and inadequate macroeconomic productivity have likewise destabilised the growth of the basin (Namara et al., 2011).

The challenges are equally observed in the riverine areas like the Niger Delta, where constant oil discharges are the main means of environmental pollution and remain a continuous receiver of unyielding burdens and attacks on its ecosystems, interrupting the well-being and living states of the people. In addition, the area is affected by water management and inadequate facilities, and the swamp in the region is regular because of the rise in sea level and other issues. In conclusion, water quality linked to health challenges such as diarrhoeal and vector-borne diseases is a major concern in the Niger basin, accompanied by respiratory diseases. These "traditional environmental risks" are enlarged by starvation and occasionally disputes. Apart from specific places, the challenges in water quality have a slight or no effect on the aquatic ecosystems and their biodiversity, which are mainly exposed by the affected land practice variation connected to the overall increase in population (Wolfram et al., 2021).

3.7.2.3 The FEWE Nexus Application

The project "Plateforme Africaine de Suivi-Assessment du Nexus Eau-Energie-Sécurité Alimentaire-Ecosystème (FEWE): cas du bassin du Niger," a partnership between JRC and the Agriculture, Hydrology and Meteorology (AGRHYMET), intended to advance the execution of the FEWE nexus method in the Niger reservoir. A complete approach was used to challenge the complicated nature of effects on the

Niger reservoir water resources by advancing an algorithm and an established situation, which unites land, climate utilisation, and administrative predictions. In addition, the AGRHYMET Regional Centre designated two model subbasins (the Sirba River basin and the multipurpose Sélingué Dam basin in the Sankarani River), bearing in mind not only the competitions connected to FEWE but also the existence of socio-economic and biophysical data. Formerly, the application of the FEWE nexus method in the Mékrou River reservoir (a branch of the Niger joint by Niger, Benin, and Burkina Faso) revealed poor food production because inadequate local manufacture could be decreased by about one-third by boosting the implementation and effective circulation of irrigation and fertilisers (Udias et al., 2018).

The challenges linked to weather fluctuations and climate change increase the growing demand for water and increased population pressure in Nigeria, one of the most densely inhabited countries on the continent. The major sustainable renewable energy within the region is hydropower, an appropriate replacement for fossil fuel, which remains the main environmental challenge in the Delta area (flaring of natural gas, oil spillages, and misuse of natural resources) and results in related effects on farming and fisheries. Socio-economically, the situation in the region is highly worsened by fluctuations in oil prices, security intimidations, and natural calamities such as flooding (Pingault et al., 2017). Given this complex scenario, data availability and accessibility are the biggest challenges because of discontinued and disjointed monitoring and incomplete funding. The case of this last challenge prompted, for instance, the Nigerian Meteorological Agency (NIMET) to the conclusion of data commercialisation. Moreover, the deficiency of required provisions for gas and oil companies working in the delta area to produce their data for research institutions as well as the government causes a huge restriction in the perspective of research in the area.

Generally, water demand is of critical importance in Southern Nigeria regarding the richness of natural resources, arranged by appropriate rainfall rates and frequency of rain-fed agriculture. Nevertheless, the region is prone to dangers associated with governance and institutional disappointment, inadequate infrastructures, scarce maintenance, and natural tragedies. Environmental pollution and ecosystem degradation, along with biodiversity conservation, are now prioritized at the top of the political agenda (Pingault et al., 2017).

3.8 SUSTAINING THE AFRICAN SCENARIO TO ACHIEVE FEWE NEXUS

In the global community, Africa has been recognised as the fastest-growing continent based on available records. The estimation is that approximately one in two people in the global population will be in Africa in the coming decades. Besides the fast growth in domestic incomes and financial activities, the demand for natural resources, including FEWE services, will be quantumly energised. In the meantime, several African countries will remain impoverished unless there is a state of emergency. Immediate policy acts to inspire investments and embrace modern and affordable technology are urgently utilised. The sustainable African scenario (SAS) opines that in order to attain the African development targets by 2020–2030, every FEWE nexus development, as well as modern primary energy supply, will rise exponentially while the utilisation of traditional bioenergy for cooking is eradicated.

Generally, the energy consumption per capita remains less than one-third of the world average. The major source of energy consumption in Africa is household activities. It is expected that the number of fans, air conditioners, and refrigerators will more than double by 2030, while the standard for minimum energy consumption will be restricted to the least efficient appliances and modern energy is coded to mitigate the impact of energy demands. In the same vein, importing inefficiently used cars that consume over 50% of fuel economy will be restricted while encouraging low-carbon industrial technologies, for example, electric arc furnace (EAF) steel making, which will record over 70% of steel products and a 75% increase in electricity demand by 2030. Also, renewable energy, especially solar PV, will account for most new capacity due to ever-declining costs driven by rapid global uptake. However, solar PV and wind technology will provide 27% of power generation by 2030, supplying eight times more electricity than the present-day generation.

To precisely model and evaluate the FEWE Nexus, it is suitable to create data that can measure FEWE flow, make numerical estimates, and predict the related costs. The Multi-Scale Integrated Assessment of Society and Ecosystem Metabolism (MuSIASEM) instrument was established to model the FEWE nexus (Simpson and Berchner, 2017). This tool demonstrates the metabolic designs of FEWE regarding socio-economic and ecological features. MuSIASEM enables the concurrent application of ecological, social, and demographic variables, even when they are described at several points and scales. In this case, it allows an efficient analysis of the relationships between FEWE at the local and international levels. Moreover, MuSIASEM integrates checks for proposed situations' viability, feasibility, and desirability.

3.8.1 Description of Policy Implications on the FEWE Nexus in Africa

Realising the objectives of SDG is one of the characteristics that ensures its aim remains the top priority in the political programme on the continent, calling for a combined effort and teamwork among all appropriate players. Growth plans and strategic formations should be carefully united with goals and targets. The recognised agenda should be modernised to be more operative in the execution of the policies, and suitable points of sponsorship should be followed in the global market. This could arise not only from the contributors but also from the participation of public–private companies. Regarding global waterways, the management of transboundary challenges remains one of the major issues in several African regions' institutional and political plans. Some years ago, important advancing steps were made, with the production of the major river basin organisations, while encouraging the relationships between the regional economic communities and the national governments.

Water mediation is the greatest and most essential instrument the organisational players can hold, including transborder water administration. Collaboration with water diplomacy and a strong science-plan discourse remains the greatest and most appropriate route to achieve sustainable development in curtailing population increase, socio-economic dynamics, and environmental changes. Encouraging relationships and teamwork among states, national, and supranational organisations, investors from several areas and local communities are significant when analysing infrastructure growth and water management declarations in a wider situation. Implementing a collaborative method would improve strong decisions and guarantee wider acceptance during the execution of these policies. In the transboundary water management background, realising a unified organisation would be meaningful by coordinating norms and data among the riparian states within a river basin and creating definite procedures for data exchange. Such an arrangement could provide a common basis for establishing deliberations and contracts.

The FEWE nexus method stands as an exclusive model for strengthening food, energy, and water security while maintaining the integrity of natural ecosystems through strategic arrangements for water resources. Integrating the FEWE nexus agenda into the arrangement and implementation of infrastructure projects, with emphasis on combining it with the cheap analysis and environmental impact evaluation points, would improve the efficiency of the policymaking process by generating pervasive information. Nevertheless, effectively applying the FEWE nexus agenda entails an in-depth assessment of the exact framework. This involves analysing precise sets of significances, involving stakeholders properly, and implementing the highly appropriate dialogue model method, all directed by recognising intervention plans to be described.

The mitigation approach, integrating sustainability and climate change adaptation, and addressing transboundary dynamics and natural resource administration within the advancement database, can be a collective and supportive method for attaining sustainable resource management and growth. In this framework, technical and scientific research seems to be a critical factor in advancing an unfathomable understanding of natural and socio-economic systems, which are becoming gradually difficult. Improving research efficiency, mostly when performed within the basis of global and local arrangements, requires the contribution of specialised think tanks encompassing academics, policymakers, and private sector representatives. Such a partnership would guarantee more knowledgeable decision-making and contribute considerably to building trust among investors who share and compete for common resources.

In this situation, prioritising the advancement of digital devices and optimum practices, pursuing standardisation tasks, and building expertise for continuously integrating water resources across the continent are crucial necessities. The distribution of digital instruments and modern technological practices must be accompanied by human capacity growth to utilise these advanced technical instruments competently. Presently, there is an increase in demand to encourage the education and training of scientists and engineers with the ability to access water and interrelated areas, mainly within the young population on the continent. This target should be enthusiastically followed by encouraging higher education and training agendas within African institutions and universities focusing on science and technology. To limit the movement of skilled labour towards higher profitable areas, efforts should be made to improve the demand in the water sector.

Human capacity development programmes should involve local performers such as river basin organisations (RBOs) and regional economic communities (RECs). Coordination and expansion of local and regional programmes are essential to address broader challenges in Africa's water, energy, food, and ecosystem sectors. Improving capabilities in the water sector requires full integration of the economic and scientific disciplines combined with expertise in water diplomacy and political science. Emphasis must be placed on understanding groundwater and surface water resources, their connections, and the vigilant surveillance of potential quality issues that could harm human health.

3.9 CONCLUSION

The current research intends to strengthen the security of Africa's food, energy, water, and ecosystems. The analysis emphasises the assessment of the FEWE nexus within different locations on the continent. Enhancing human health and incomes are the key challenges identified within the region. Nevertheless, challenges arise due to Africa's high susceptibility to weather inconsistency, poverty reduction, guaranteeing food, energy, and water security, natural climate modification, and sustainable river basin management. The intricacy of these problems across every length and breadth in Africa hinders the utilisation of every part of the strategy outlooks that encourage the implementation of broad agendas which study their several collaborations. The 2063 Agenda of the African Union titled "The Africa We Want" predicts an affluent Africa focusing on comprehensive development and sustainable growth. In order to achieve these goals, food, energy, and water security must be guaranteed through the advancement and functionalisation of distributive devices at the essential scale. Given this, the FEWE nexus methods could manage and govern several sectors involved, identifying their mutuality and the cost of natural assets, promising harmonisation among investors, and recognising appropriate procedural resolution to enhance trade-offs and exploit synergies across sectors. However, concerted efforts and regional collaborations through the FEWE nexus can reinforce less-developed countries, strengthen those already established, protect the environment, and guarantee security across all crucial resources for the coming generations in Africa. At its core, collaboration improves security in food, energy, water, and ecosystems while securing ecosystems. This method fortifies support and fosters corporations across different stages of governance, enabling the collective use of resources. These collective efforts help reduce inequalities and alleviate poverty by expanding individual countries' access to resources, thus developing considerable socio-economic growth.

REFERENCES

'Africa Energy Outlook 2022' (2022) *Africa Energy Outlook 2022*. Doi: 10.1787/2abd9ce2-en

Alfieri, L. et al. (n.d.) 'Global projections of river flood risk in a Warmer world'. *Earth's Future*, 5(2), pp. 171–182.

Aliyu, A. K., Modu, B. and Tan, C. W. (2018) 'A review of renewable energy development in Africa: A focus in South Africa, Egypt and Nigeria'. *Renewable and Sustainable Energy Reviews*, 81, pp. 2502–2518.

Apeh, O. O. and Nwulu, N. I., (2025). Improving traceability and sustainability in the agri-food industry through blockchain technology: A bibliometric approach, benefits and challenges. *Energy Nexus*, 17, p. 100388.

Apeh, O. O., Overen, O. K. and Meyer, E. L. (2021) 'Monthly, seasonal and yearly Assessments of global solar radiation, clearness index and diffuse fractions in Alice, South Africa'. *Sustainability (Switzerland)*, 13(4), pp. 1–15. Doi: 10.3390/su13042135

Arias, M. E., Farinosi, F. and Hughes, D. A. (2022) 'Future hydropower operations in the Zambezi River basin: Climate impacts and adaptation capacity'. *River Research and Applications*, 38(5), pp. 926–938.

Atheru, Z. (2018) 'The African networks of centres of excellence on water sciences PHASE II (ACE WATER 2)'. *IGAD Climate Prediction and Application Centre (ICPAC).* https://aquaknow.jrc.ec.europa.eu/system/files/JRC_Technical%20Report%20ACE2-HCDAchievements.pdf

Aude, N. et al. (2017) *Science for the AU-EU partnership-building knowledge for sustainable development.* European Commission.

Bejarano, M. D., Jansson, R. and Nilsson, C. (2018) 'The effects of hydropeaking on riverine plants: A review'. *Biological Reviews*, 93(1), pp. 658–673.

Belinskij, A. (2015) 'Water-energy-food nexus within the framework of international water law'. *Water*, 7(10), pp. 5396–5415.

Birol, F. et al. (2014) *World energy outlook 2014.* http://www.oecd-ilibrary.org/energy/world-energy-outlook-2014_weo-2014-en

Botai, J. O. et al. (2021) 'A review of the water–energy–food nexus research in Africa'. *Sustainability*, 13(4), p. 1762.

Buhaug, H. and Vestby, J. (2019) 'On growth projections in the shared socioeconomic pathways'. *Global Environmental Politics*, 19(4), pp. 118–132.

Busch, S. (n.d.) *Analysis of the water-power nexus in the Southern African power pool.* Doi: 10.2760/920794

Busch, S., De Felice, M. and González, I. H. (2020) *Analysis of the water-power nexus in The Southern African power pool.* European Union.

Busker, T. et al. (2019) 'A global lake and reservoir volume analysis using a surface water dataset and satellite altimetry'. *Hydrology and Earth System Sciences*, 23(2), pp. 669–690.

Camposeco-Negrete, C. (2013) 'Optimization of cutting parameters for minimizing energy consumption in turning of AISI 6061 T6 using Taguchi methodology and ANOVA'. *Journal of Cleaner Production*, 53, pp. 195–203.

Chaturvedi, A. K. (2013) *Water: A source for future conflicts.* Vij Books India Pvt Ltd.

Commission, A. U. (2021) *Africa regional overview of food security and nutrition 2020: Transforming food systems for affordable healthy diets.* Food & Agriculture Org.

Conway, D. et al. (2017) 'Hydropower plans in eastern and southern Africa increase risk of concurrent climate-related electricity supply disruption'. *Nature Energy*, 2(12), pp. 946–953.

da Silva, R. C., de Marchi Neto, I. and Seifert, S. S. (2016) 'Electricity supply security and the future role of renewable energy sources in Brazil'. *Renewable and Sustainable Energy Reviews*, 59, pp. 328–341.

Damkjaer, S. and Taylor, R. (2017) 'The measurement of water scarcity: Defining a meaningful indicator'. *Ambio*, 46(5), pp. 513–531.

Danert, K., Adekile, D. and Canuto, J. G. (2020) 'Striving for borehole drilling professionalism in Africa: A review of a 16-year initiative through the rural water supply network from 2004 to 2020'. *Water*, 12(12), p. 3305.

Dargin, J., Daher, B. and Mohtar, R. H. (2019) 'Complexity versus simplicity in water energy food nexus (WEF) assessment tools'. *Science of the Total Environment*, 650, pp. 1566–1575.

De Felice, M., González Aparicio, I., Huld, T., Busch, S., & Hidalgo González, I. (2019). *Analysis of the water-power nexus in the West African Power Pool* (JRC Technical Report No. JRC115157). Publications Office of the European Union. https://doi.org/10.2760/362802

Dobaa, A. M., Abbas, M. S. and Metwaly, K. S. (2014) 'Impacts of biofuel on environment in some African countries'. In *International Conference of Water and Energy in Nile Basin Countries – Potentialities of Development and Integration*, Cairo, Egypt.

Dosio, A. (2017) 'Projection of temperature and heat waves for Africa with an ensemble of CORDEX regional climate models'. *Climate Dynamics*, 49(1–2), pp. 493–519.

Dosio, A. et al. (2018) 'Extreme heat waves under 1.5 °C and 2 °C global warming'. *Environmental Research Letters*, 13(5), p. 54006.

Dosio, A. et al. (2019) 'What can we know about future precipitation in Africa? Robustness, Significance and added value of projections from a large ensemble of regional climate models'. *Climate Dynamics*, 53, pp. 5833–5858.

Farinosi, F. et al. (2017) *Towards an African Atlas of water energy food cooperation–data collection*. JRC Technical Report no. JRC108695. Under Editing and Publishing Process.

Farinosi, F. et al. (2018) 'An innovative approach to the assessment of hydro-political risk: A spatially explicit, data driven indicator of hydro-political issues'. *Global Environmental Change*, 52, pp. 286–313.

Foong, A. et al. (2022) 'Adjusting agricultural emissions for trade matters for climate change mitigation'. *Nature Communications*, 13(1), p. 3024.

Frieler, K. et al. (2017) 'Assessing the impacts of 1.5 C global warming–simulation protocol of the inter-sectoral impact model intercomparison project (ISIMIP2b)'. *Geoscientific Model Development*, 10(12), pp. 4321–4345.

Gabr, M. E. (2022) 'Modelling net irrigation water requirements using FAO-CROPWAT 8.0 and CLIMWAT 2.0: A case study of Tina plain and East South ElKantara regions, North Sinai, Egypt'. *Archives of Agronomy and Soil Science*, 68(10), pp. 1322–1337.

Gebreyohannes, G. and Zenebe, G. (2019) 'Climate-smart energy use (management)'. In Kiros Meles Hadgu (Ed.), *Climate-smart agriculture: Enhancing resilient agricultural systems, landscapes, and livelihoods in Ethiopia and beyond* (pp. 141–155). World Agroforestry Centre.

Hajjar, B. (2020) *The children's continent: Keeping up with Africa's growth*. World Economic Forum.

Hirwa, H. et al. (2021) 'Insights on water and climate change in the greater horn of Africa: Connecting virtual water and water-energy-food-biodiversity-health nexus'. *Sustainability*, 13(11), p. 6483.

Hirwa, H. et al. (2022) 'Water accounting and productivity analysis to improve water savings of Nile River Basin, East Africa: From accountability to sustainability'. *Agronomy*, 12(4), p. 818.

Hoegh-Guldberg, O. et al. (2019) 'The human imperative of stabilizing global climate change at 1.5° C'. *Science*, 365(6459), p. eaaw6974.

Hogeboom, R. J., Knook, L. and Hoekstra, A. Y. (2018) 'The blue water footprint of the world's artificial reservoirs for hydroelectricity, irrigation, residential and industrial water supply, flood protection, fishing and recreation'. *Advances in Water Resources*, 113, pp. 285–294.

Hughes, D. A. and Farinosi, F. (2020) 'Assessing development and climate variability impacts on water resources in the Zambezi River basin. Simulating future climate and development scenarios'. *Journal of Hydrology: Regional Studies*, 32, p. 100763.

Hughes, D. A. and Farinosi, F. (2021) 'Unpacking some of the linkages between uncertainties in observational data and the simulation of different hydrological processes using the Pitman model in the data-scarce Zambezi River basin'. *Hydrological Processes*, 35(4), p. e14141.

International Renewable Energy Agency. (2015). *REmap 2030: A renewable energy roadmap*. Abu Dhabi: Author. https://www.irena.org/-/media/Files/IRENA/Agency/Publication/2015/IRENA-_REthinking_Energy_2nd_report_2015.pdf

IRENA. (2017) 'Rethinking energy 2017: Accelerating the global energy transformation'. *International Renewable Energy Agency*. Doi: 10.1517/13543780902854194

Jemmali, H. (2018) 'Water poverty in Africa: A review and synthesis of issues, potentials, and policy implications'. *Social Indicators Research*, 136, pp. 335–358.

Kanda, E. K. et al. (2023) 'Water-energy-food-ecosystem nexus and sustainable development in the horn of Africa'. *F1000Research*, 12(143), p. 143.

Lehner, B. et al. (2011) 'High-resolution mapping of the world's reservoirs and dams for sustainable river-flow management'. *Frontiers in Ecology and the Environment*, 9(9), pp. 494–502.

Lin, D., Wambersie, L., & Wackernagel, M. (2021). Estimating the date of Earth Overshoot Day 2022. *Global Footprint Network*. https://www.overshootday.org/content/uploads/2022/06/Earth-Overshoot-Day-2022-Nowcast-Report.pdf

Louhichi, K. et al. (2020) *Modelling farm-household livelihoods in developing economies*. EUR.

Lynd, L. R. et al. (2015) 'Bioenergy and African transformation'. *Biotechnology for Biofuels*, 8(1), pp. 1–18.

Macknick, J. et al. (2012) 'Operational water consumption and withdrawal factors for electricity generating technologies: A review of existing literature'. *Environmental Research Letters*, 7(4), p. 45802.

Markantonis, V. et al. (2019) 'Can the implementation of the water-energy-food nexus support economic growth in the Mediterranean region? The current status and the way forward'. *Frontiers in Environmental Science*, 7, p. 84.

Marti, L. and Puertas, R. (2020) 'Analysis of the efficiency of African countries through their ecological footprint and biocapacity'. *Science of the Total Environment*, 722, p. 137504.

Medinilla, A. (2017). *Understanding the Niger Basin Authority (NBA/ABN): Reconciling upstream and downstream interests on the Niger River*. European Centre for Development Policy Management (ECDPM). Retrieved from https://ecdpm.org/download_file/1832/1317

Mekonnen, M. M. and Hoekstra, A. Y. (2012) 'The blue water footprint of electricity from hydropower'. *Hydrology and Earth System Sciences*, 16(1), pp. 179–187.

Meyer, E. L., Apeh, O. O. and Overen, O. K. (2020) 'Electrical and meteorological data acquisition system of a commercial and domestic microgrid for monitoring PV parameters'. *Applied Sciences (Switzerland)*, 10(24), pp. 1–18. Doi: 10.3390/app10249092

Migali, S., et al. (2018). *International Migration Drivers: A Quantitative Assessment of the Structural Factors Shaping Migration*. Joint Research Centre. Retrieved from https://publications.jrc.ec.europa.eu/repository/handle/JRC112622

Miller, C. A. and Wyborn, C. (2020) 'Co-production in global sustainability: Histories and theories'. *Environmental Science & Policy*, 113, pp. 88–95.

Muhirwa, F. et al. (2022) 'Ecological balance emerges in implementing the water-energy-food security nexus in well-developed countries in Africa'. *Science of the Total Environment*, 833, p. 154999.

Muhirwa, F. et al. (2023) 'Alert in the dynamics of water-energy-food production in African countries from a nexus perspective'. *Resources, Conservation and Recycling*, 194, p. 106990.

Muraina, T. O. et al. (2023) 'Grassland cover declined in Southern Africa but increased in other African subcontinents in the early twenty-first century'. *Environmental Monitoring and Assessment*, 195(5), p. 621.

Namara, R. E. et al. (2011) *An overview of the development challenges and constraints of the Niger Basin and possible intervention strategies*. IWMI.

Naumann, G. et al. (2018) 'Global changes in drought conditions under different levels of warming'. *Geophysical Research Letters*, 45(7), pp. 3285–3296.

Niasse, M. and Varis, O. (2021) 'Quenching the thirst of rapidly growing and water-insecure cities in sub-Saharan Africa'. In Cecilia Tortajada & Eduardo Araral (Eds.), *Global water resources* (pp. 283–305). Routledge.

Nicholson, S. E., Funk, C. and Fink, A. H. (2018) 'Rainfall over the African continent from the 19th through the 21st century'. *Global and Planetary Change*, 165, pp. 114–127.
OECD. (2018) *Nations AOotU*. OECD-FAO Agricultural Outlook.
Orimoloye, I. R. (2022) 'Water, energy and food nexus: Policy relevance and challenges'. *Frontiers in Sustainable Food Systems*, 5, p. 824322.
Pavičević, M. et al. (2020) *Analysis of the water-power nexus in the North, Eastern and Central African power pools*. Publications Office of the European Union.
Pavičević, M. et al. (2021) 'Water-energy nexus in African power pools–The Dispa-SET Africa model'. *Energy*, 228, p. 120623.
Pingault, N. et al. (2017) *Nutrition and food systems. A report by the High high-level panel of experts on food security and nutrition of the committee on world food security*. FAO.
Putra, M. P. I. F., Pradhan, P. and Kropp, J. P. (2020) 'A systematic analysis of water-energy-Food security nexus: A South Asian case study'. *Science of the Total Environment*, 728, p. 138451.
Raftery, A. E. et al. (2017) 'Less than 2 C warming by 2100 unlikely'. *Nature Climate Change*, 7(9), pp. 637–641.
Ramos, M. P. et al. (2020) *Assessing market incentive policies in Kenya with a food security and nutrition perspective: A macro-microsimulation approach*. Publications Office of the European Union.
Ranzato, M. et al. (2018) *Co-production and the environment (2018) co-production and co-creation* (pp. 180–190). Taylor and Francis.
Riahi, K. et al. (2017) 'The shared socioeconomic pathways and their energy, land use, and greenhouse gas emissions implications: An overview'. *Global Environmental Change*, 42, pp. 153–168.
Rodriguez, D. J. et al. (2013) *Thirsty energy*. World Bank.
Sadc, S., Zamcom, G. and Arendal, U. (2012) *Zambezi River Basin Atlas of the changing environment*. Gaberone, Harare and Arendal.
Sall, M. et al. (2020) 'Water constraints and flood-recession agriculture in the Senegal River valley'. *Atmosphere*, 11(11), p. 1192.
Sanchez, R. G. et al. (2020) 'Freshwater use of the energy sector in Africa'. *Applied Energy*, 270, p. 115171.
Shumilova, O. et al. (2018) 'Global water transfer megaprojects: A potential solution for the water-food-energy nexus?' *Frontiers in Environmental Science*, 6, p. 150.
Simpson, G. and Berchner, M. (2017) 'Water-energy nexus-measuring integration: Towards a water-energy-food nexus index'. *Water Wheel*, 16(1), pp. 22–23.
Spalding-Fecher, R., Joyce, B. and Winkler, H. (2017) 'Climate change and hydropower in the Southern African power pool and Zambezi River Basin: System-wide impacts and policy implications'. *Energy Policy*, 103, pp. 84–97.
Spinoni, J. et al. (2020) 'Future global meteorological drought hot spots: A study based on cordex data'. *Journal of Climate*, 33(9), pp. 3635–3661.
Tashtoush, F. M., Al-Zubari, W. K. and Shah, A. (2019) 'A review of the water-energy-food nexus measurement and management approach'. *International Journal of Energy and Water Resources*, 3, pp. 361–374.
Twinoburyo, E. N. et al. (2019) *Africa 2030: Sustainable development goals three-year reality check*. Sustainable Development Goal Center for Africa (SDGC/A).
Udias, A. et al. (2018) 'A decision support tool to enhance agricultural growth in the Mékrou river basin (West Africa)'. *Computers and Electronics in Agriculture*, 154, pp. 467–481.
Union, A. (2014) *Malabo declaration on accelerated agricultural growth and transformation for shared prosperity and improved livelihoods doc. Assembly/Au/2 (Xxiii)*. African Union.
Vaucelle, S. (2015) 'Le fleuve Niger et son bassin: Aménagements, gouvernance et stratégies d'adaptation au changement climatique'. *Les Cahiers d'Outre-Mer*, 270, pp. 243–270.

Wieben, E. (2016) *The post-2015 development agenda: How food loss and waste (FLW) reduction can contribute towards environmental sustainability and achieving the sustainable development goals*. United Nations University Institute for Integrated Management of Material.

Wolfram, J. et al. (2021) 'Water quality and ecological risks in European surface waters–monitoring improves while water quality decreases'. *Environment International*, 152, p. 106479.

Yalew, A. W. (2022) 'The Ethiopian energy sector and its implications for the SDGs and modelling'. *Renewable and Sustainable Energy Transition*, 2, p. 100018.

Zar, C. et al. (2015) 'A global boom in hydropower dam construction'. *Aquatic Sciences*, 77(1), pp. 161–170.

Zhang, J. et al. (2022) 'Mapping the complexity of the food-energy-water nexus from the lens of sustainable development goals in China, *Resources, Conservation and Recycling*, 183, p. 106357.

Zwarts, L. et al. (2006) 'The economic and ecological effects of water management choices in the Upper Niger river: Development of decision support methods'. *Water Resources Development*, 22(1), pp. 135–156.

4 Towards Societal-Inclusive Sustainability

Developing Interdisciplinary Strategies in FEW Nexus Research and Practice

4.1 INTRODUCTION

A significant fraction of the world population still faces inadequate access to FEW resources. Despite the efforts to reach the Millennium Development Goals (MDGs) established by the UN over the last two decades, roughly 800 million population continue to experience food insecurity, about 800 million do not have access to safe drinking water, and approximately 1.2 billion lack access to electricity (FAO et al., 2015; United Nations, 2015). The predicted increase in world population from 7.4 billion in 2016 to 9.7 billion in 2050 (UN ESCAP, 2017) is accompanied by a 1–5% increase in per capita GDP in many countries (Hawksworth and Chan, 2015). Also, with diet modification and the consumption of more meat protein, there is an expected significant increase in demand and stress on available FEW resources. This relationship of FEW resources is regularly described using statements such as "feeding ourselves thirsty" (Rodriguez et al., 2013; Barton, 2015).

The complicated interdependences worsen the pressure exerted upon the limited FEW resources. Globally, food production needs a substantial quantity of water, representing around 70% of total freshwater withdrawal and 90% of freshwater consumption (Pena-Arancibia et al., 2020). On the other hand, food production is less energy-intensive, accounting for about 30% of world energy consumption (Schramski, Woodson and Brown, 2020). Nevertheless, an expected 25–30% of world food production is lost during processing periods, post-harvest, or misused at consumer stages or retail, resulting in substantial waste in both energy and water resources (Kummu et al., 2012). Although food is a source of energy, the utilisation of food crops for biofuels has attracted much attention to the high-water concentration of crop production, as reflected in phrases, for example, "blue impacts of green energy" or "cars versus carnivores" (Mathioudakis et al., 2017). However, an outline of the few nexus security issues is presented in Figure 4.1. It mostly centred on the connections between FEW security nexus resources and the benefits society could derive.

Energy production requires significantly less water than food production, constituting about 15% of world water extraction, with consumption constituting only 11% of this water (Walton, 2018). Water is utilised for oil and gas withdrawal and

DOI: 10.1201/9781003633846-4

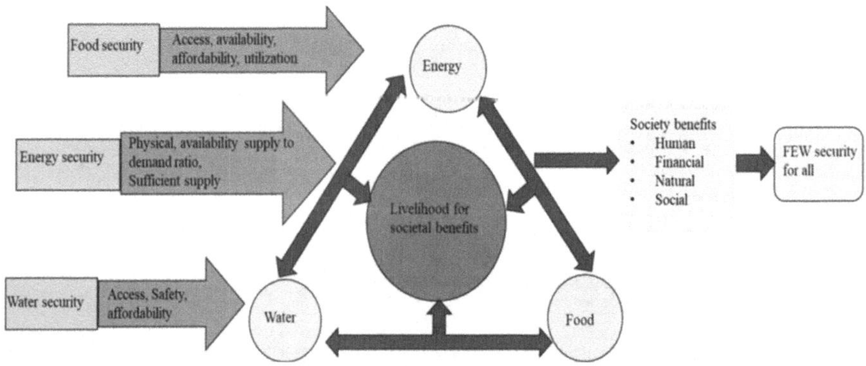

FIGURE 4.1 The outline of the FEW nexus for societal livelihoods.

electricity generation (a secondary form of energy), mostly through thermoelectric and hydroelectricity cooling in nuclear power plants and fossil fuels (Healy et al., 2015). There is continuous deliberation regarding the high-water intensity of oil and gas withdrawal from alternative low-permeability basins in the United States through horizontal drilling and hydraulic fracturing. However, research shows these processes account for less than 1% of freshwater extraction nationally. The fraction can be much higher locally (Chen and Carter, 2016). With the possibility of growing climate excesses, including more intense floods and longer-term droughts, collisions among FEW resources are expected to become more common (Apeh and Nwulu 2024). The example is pronounced at the 2003 European heat wave, a disastrous collision between energy and water, contributing to an estimated 70,000 deaths, partly due to nuclear power plant cessations caused by low river levels and higher water temperatures restricting cooling processes (Alfieri et al., 2018). Diminished access to FEW resources can lead to social unrest and affect global security (Willis et al., 2016).

A study of the interconnectedness of FEW resources and their effect on the Earth system is crucial for protecting the accessibility of reasonable and consistent resources in an environmentally sustainable manner. Several new scientific agendas have been established to discuss these nexus challenges within the framework of Earth and the climate system. Examples consist of the Belmont Forum's Sustainable Urban Global Initiative. The US National Science Foundation launched the Innovations at the Nexus of Food, Energy and Water Systems (INFEWS) programme, aimed at supporting research to comprehend existing and anticipated intertwined procedures connected to the environment and society. This programme covers biological processes such as agroecosystem productivity, physical processes including technologies for higher effective resource utilisation, and natural processes such as hydrology. Figure 4.2 presents the recommendations that provide a basis for interdisciplinary research advancement in the FEW nexus.

There is growing importance in behavioural and social processes associated with governance and policymaking. Computational, monitoring, networking methods, and visualisation are crucial for evaluation and policymaking. The US Department of Energy likewise maintains a healthy package centred on the energy and water nexus (Heidari and Khovalyg, 2020).

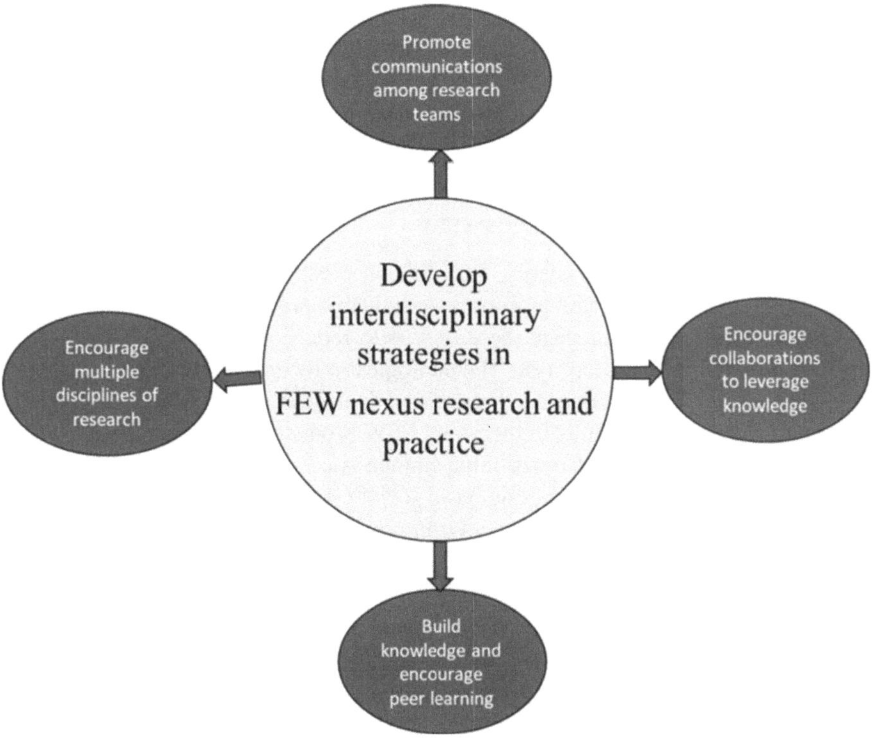

FIGURE 4.2 The essential set of recommendations for promoting interdisciplinary research and practices.

4.2 THE PERCEPTION OF FEW NEXUS FROM DIVERSE STUDY AREAS

Over the past decade, the FEW nexus has continued to generate increasing consideration. However, the rising volume of research combined with variable explanations of the FEW nexus ideas can be unclear for novices. This section will unravel the essential approaches that are employed to examine and understand the nexus: 1) the ecosystem's perception, 2) the waste management perception, 3) the institutional change perception, 4) the stakeholder trust perception, and 5) the learning process perception, which is presented in Figure 4.3.

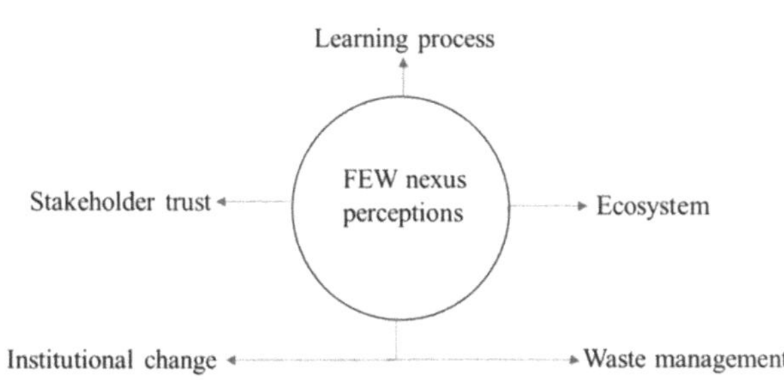

FIGURE 4.3 The perceptions of the FEW nexus across different fields.

It is worth noticing that these perceptions are not entirely different and do not include all possible ways the FEW nexus is studied. In particular, they do not discuss the critical work on technological solutions like biofuels (Hassanpour Adeh, Selker and Higgins, 2018). Nevertheless, these perceptions offer valuable understandings of all-embracing principles that affect FEW nexus research.

4.2.1 The Perception from Ecosystems

From the viewpoint of ecosystems, a significant emphasis is placed on measuring the environmental impacts and promoting sustainable practices within FEW sectors. The main objective is to manage these sectors to reduce environmental destruction while maintaining FEW securities. The ecological effect serves as a critical restriction. If an action exceeds a definite threshold of damage to the environment, it should be withdrawn, even if it results in increased FEW production. This perception aligns with establishing a safe and functioning human space (Raworth, 2017). This idea addresses the significance of meeting basic social needs without surpassing ecological limits and producing undesirable environmental harm. For example, in arid areas with water shortages, drawing large amounts of water from rivers may increase crop yield but may cause decreased stream flow and a high level of fish mortality beyond the acceptable level for that area. Thus, in this situation, although drawing more water could increase food production, the activity should be halted because of its environmental consequences.

A real-world illustration of this problem was the weak application of fossil groundwater in Saudi Arabia for wheat cultivation, which at one point made Saudi Arabia the 10th largest wheat producer globally (DeNicola et al., 2015). On the other hand, rather than entirely assessing damage, some researchers have endeavoured to measure the advantages humanity derives from ecosystems, often termed "Ecosystem services." Several of these research studies integrate the ecosystem as an essential element of the nexus, classifying it as the FEWE nexus (or variations thereof) (Fasel et al., 2016).

One of the most prevalent devices for evaluating the environmental effects of products and processes is the Life Cycle Assessment (LCA) (Tabatabaie, Bolte and Murthy, 2018). LCA is a standardised methodology for measuring ecological effects, encompassing four stages: defining goals and scope, compiling data, conducting a life cycle impact assessment, and interpreting results (Walsh, Murray and O'Sullivan, 2015). In LCA, each system is broken down into unit processes, and for each unit process, material flows, resource consumption, and energy use are calculated and combined to signify the environmental effects of that system (Mannan et al., 2018). There is another approach to measuring environmental effects, which involves the use of ecological indicators. This technique includes a set of indicators recognised to denote sustainability in an exact location, and the standards of these indicators are then compared against those of other locations or established standards. Some of these indicators involve land use, fertiliser consumption, population overweight, added value from agriculture, and several others (Saladini et al., 2018).

The different means for measuring environmental effects differ in their efficiency and applicability. LCA methods are often adept at examining impacts on a local or regional scale but may not constantly include them to account for global limitations. Furthermore, while LCA's attention to ecosystem services is critical, measuring them can be challenging. The most appropriate device will often centre on the exact scope of inquiry that specific researchers are pursuing. The ecosystem's perception addresses the significance of assessing possible environmental effects associated with the FEW nexus in every policymaking process. Within this perception, LCA and ecological indicators appear to be the most typically employed devices when investigating the nexus.

4.2.2 The Perception of Waste Management

The perception from waste management addresses the key aspect of waste in protecting resource security, mainly in food, energy, and water. Encouraging supply will not assist; a decrease in resource demand is likewise important. All three elements of the FEW nexus are highly connected to food production. Hence, when food is wasted, it involves the wastage of all the resources used in its cultivation, harvesting, treatment, processing, and transportation. When one thinks about the inputs (such as energy, water, and nutrients) needed for each material contained in food production (such as pesticides, herbicides, fertilisers, agricultural machinery, and fuel), it is apparent that each part of wasted food represents a considerable source of inadequacy within the nexus (Hannibal and Vedlitz, 2018). Yearly, a considerable quantity of food is lost at different phases of the food supply chain. According to the FAO, virtually one-third of globally produced food is wasted or lost, amounting to about 1.3 billion tons annually (Gustavsson et al., 2011).

While the overall stages of food losses in developing and advanced countries are the same, the levels at which these wastes occur vary. For example, in a developing country, over 40% of food waste happens post-harvest and during processing, while in advanced nations, over 40% of waste arises at the consumer and retail levels. Water is misused directly through unnecessary irrigation of crops and in domestic locations such as grass watering and unnecessary household usage. Inefficiencies in water applications likewise cause an increase in nutrient wastes, causing adverse ecological effects. Energy is wasted through ineffective transmission lines and the use of ineffective vehicles such as cars, trucks, and other means of energy transport. While there is consensus on the implications of waste management, eradicating waste stances is a daunting challenge, requiring complete knowledge of world supply chains, socio-economic factors, and human behaviour.

One progressively dominant agenda for knowing waste management is the idea of the "Circular Economy," described by Geissdoerfer et al. (2017) as "a reformative system that reduces resource input and waste, discharges, and energy leakage by reducing, closing, and contracting material and energy loops." This system differs from the traditional "linear" economies that generate objects, consume them, and produce considerable waste. Promoters of the circular economy state that it can be accomplished by prioritising robust design and the reuse and recycling of materials. This perception has gathered support from several governmental bodies, such as China's "Circular Economy Promotion Law of the People's Republic of China" and the European Union's (EU) "Circular Economy Strategy" (Geissdoerfer et al., 2017).

Moreover, several research works have been published to investigate the relationship between a circular economy and the FEW nexus (Del Borghi, Moreschi and Gallo, 2020). The waste management viewpoint investigates the effect of nutrient retrieval and waste decrease on the FEW nexus. Rather than concentrating exclusively on growing the supply of FEW resources, waste management emphasises decreasing demand and enhancing waste recycling. Several governments have already established strategies to promote waste management to progress towards a "circular economy."

4.2.3 The Perception of Institutional Change

The perception of institutional change centres on the activities of institutions, inspecting how policies have formerly affected and are presently influencing the FEW nexus. Similarly, it seeks to determine the tactics for using this knowledge to produce functional policies. At its principal stage, the FEW nexus attempts to develop a holistic study of systems, exceeding isolated opinions of individual resources. This implies that institutions, traditionally managed in isolation, must now create new connections between sectors and generate fresh avenues for collaboration. Realising this demands time, political determination, and financial resources. Research from this perception constitutes a significant and varied part of FEW nexus studies. The nexus is often presented as a scope for policymaking, and research from the Institutional Change perception directly investigates this possibility.

Before exploring definitive research, it is worth observing the political and financial sustenance sources for the significant part of the work within the institutional perception. Several multinational and government parastatals have shown strong interest in the FEW nexus idea over the last decade. The US National Science Foundation (NSF) collaborated with the US Department of Agriculture (USDA) to sponsor the ongoing INFEWS funding, which allocates millions for FEW nexus research annually. China's NSF has introduced FEW initiatives to hasten research in FEW, aligning with a circular economy strategy (Rodríguez et al., 2019). The concept of nexus has been integrated both implicitly and explicitly in Europe into different reports presented by the Energy Strategy, Water Framework Directive (WFD), and the EU's Common Agricultural Policy (CAP) (Venghaus and Hake, 2018). Identifying the nexus has likewise been emphasised for accomplishing the UN SDGs, mostly Goals 2, 6, and 7, which relate to food, water, and energy, respectively (United Nations General Assembly, 2015). The United Nations Commission for Europe (UNECE) 2018 issued an exact methodology for estimating the FEWE nexus in transboundary basins (De Strasser et al., 2016). Powerful top-down organisations support the nexus, boosting the production of an increasing team of researchers and discovering the diverse connections between the FEW nexus and the institutions.

An apparent trend in the literature is that the efficiency of the FEW nexus policy application is highly inspired by the socio-political environment of the areas where the policies are passed and the basic power dynamics are associated with the resources in every place. Thus, several patterns have been employed to estimate the influence of the FEW nexus. Due to the moderately recent appearance of the FEW nexus idea, various papers from this perception are progressive.

Several studies concentrating on institutional change perception begin with a quantitative analysis, including input–output analysis (White et al., 2017), which is used as a base for policy recommendations. Others employ an agent-based modelling method (Yang et al., 2018) to forecast how investors and ecosystems might respond to variations in FEW nexus policies.

Another subgroup of researchers implements a less quantitative method, seeking to comprehend investor behaviours towards and insights into the nexus through surveys and interviews (Halbe et al., 2018), as well as using systematic approaches like Fuzzy Cognitive Mapping (Ghodsvali, Krishnamurthy and de Vries, 2019) to determine possible consequences for FEW nexus policies. In this situation, there is an intersection between Institutional Change and Trust perceptions, as citizen behaviours can affect policy development. Weitz et al. (2017) researched some of the dominant methods within the institutional change perception and recognised governance gaps within the nexus. The perception of institutional change concentrates on the effect of FEW nexus research on the policymaking processes of governments and other top-down organisations. Presently, studies from this perception tend to be progressive, lacking complete assessments of the outcomes of policies directly influenced by the FEW nexus. The wide range of methods emphasises the essence of FEW nexus dynamics within local politics and power dynamics.

4.2.4 The Perception of Trust

Understanding the implications and effects of the FEW nexus comprises recognising the often-ignored connection between academic theory and practical acceptance and the implementation of these concepts. While the previous section addressed the nexus's connection with major institutions, it is equally important to consider how citizens participate in it. A complete grip of investor perception sheds light on the potential support that nexus initiatives may gather.

One common technique utilised to measure people's opinions on the nexus is surveys. Whereas definite methodologies may differ, stakeholder investigation contexts intend to reveal the interaction between sectors, performers, and policy advancements in FEW nexus governance, all from the stakeholders' perception. For example, in 2015, in the United States, Portney et al. (2018) performed a national public interview survey to investigate public awareness and thoughts regarding the FEW nexus. The analysis comprised questions to evaluate people's knowledge of the water-energy nexus, water-food nexus, food-energy nexus, and correlated policies. The study discovered statistically essential learning for individuals to be familiar with the associations and consequences across all three areas. Particularly, cognizance of one nexus connection linked with knowledge of the others and vice versa. Another stakeholder investigation was carried out in Phoenix, Arizona, by White et al. (2017), using groups and individual surveys to attract stakeholder opinions and create theoretical plans to support policymakers in grasping the nexus. While these research results may not be generally appropriate, they emphasise the importance of regarding public trust in the nexus.

Moving outside straight trust measurements, there is a proposal that the present top-down approach to the nexus could be beneficial by combining extra human factors.

Middleton et al. (2015) opine that for the FEW nexus to efficiently contribute to poverty alleviation and sustainable development, it must directly discourse those who stand to gain or lose the most from FEW nexus trade-offs on an ecological fairness basis. In the same way, conducted an FEW nexus agenda that integrates sustainable livelihoods, declaring that resource security relies not only on resource availability but also accessibility and that considering livelihoods can improve the nexus's usefulness in backing up the UN's SDGs.

Another critical part concerning the FEW nexus and stakeholder trust is that virtually all the studies deliberated begin with the statement that the FEW nexus is the most appropriate basis for recognising trade-offs and synergies. Cairns and Krzywoszynska (2016) challenge this postulation, signifying that the FEW nexus is a "buzzword"—a word lacking exact meaning, with a strong confidence in what it is intended to accomplish. The article employs stakeholder assessment interviews and qualitative evaluations of the literature to describe the unclear nature of the statement, permitting it to be used in different backgrounds. This doubt can make the idea concurrently indisputable and confusing, as most defendants in the research decided with the fundamental knowledge of the nexus but found it difficult to express its precise details.

Foran (2015) asserted that the FEW nexus is still in its "early stages," and this declaration is similar to Cairns and Krzywoszynska (2016), who stress the significance of seriousness in social science combined with quantitative systems thinking analyses (Foran, 2015; Cairns and Krzywoszynska, 2016). Foran contends that the FEW nexus should be recognised within a multilevel framework, with several opinions, rules, and disagreements at every level (Foran, 2015). The dispute is that a further and detailed social science analysis could aid in the growth of the notion, allowing an improved knowledge of the fundamental drives and power dynamics that may be ignored by characteristic quantitative analysis. This, in turn, can support nexus policies in correctly recognising the beneficiaries and those who may be poorly affected by planned changes, allowing for more informed decision-making. The trust perception investigates public views on the nexus through direct contact with stakeholders and in efforts to estimate how ideas may progress in the future. Much of the existing research work from this perception promotes a more robust analysis of the meaning of the FEW nexus from a social science viewpoint and a more comprehensive integration of human elements into nexus considerations.

4.2.5 The Perception of the Learning Process

As described in Section 4.2.4, the stakeholders' behaviours and policymakers' decisions are crucial to the growth and application of the FEW nexus. Identifying that stakeholders' stands and opinions are not fixed, knowing how these stakeholders pilot complicated issues is vital and has become increasingly emphasised in the literature. A study of the FEW nexus often deals with interruptions that may be unparalleled for a certain community. The community's resilience is improved when it can efficiently study and acclimate to these changes. Assessing the learning process of stakeholders, including why and how they learn, has been discovered from different perceptions. Politically and economically, stakeholders learn through experience and their perceptions of others' actions. On the other hand, a psychological viewpoint emphasises individual stakeholders' inner mental developments and inspirations. Both viewpoints recognise that learning comprises implementing behavioural and cultural characteristics across generations (Henry, 2009).

Giampietro discovers how stakeholders learn and participate with the nexus through a semiotic process, where semiotics represents the philosophical study of signs and their explanations (Giampietro et al., 2013). According to Peirce's description of semiotics, it establishes transmittable information as a constant loop relating three nodes: (1) the interpretant—the decision-maker, analyst, and storyteller, (2) the object—the interactions analysed by the interpretant, and (3) the sign—the illustrations of the object utilised by the interpretant (data, models) (Giampietro et al., 2013). In this situation, the general public and governing bodies behave as the interpretant, the outside world (material flows related to biogeomorphological systems, demand for FEW nexus resources, etc.) as the object, and scientific inquiry as the sign. Each of these nodes impacts and is impacted by the others. The outside world and the knowledge generated by the scientific community regarding the globe affect the policymaking and story creation of governing agencies, whose policies form the world and scientific analysis. This clarification stresses the effect of culture and narrative in the growth of the nexus. Thus far, much of the emphasis in nexus research has been on quantitative understanding, the "sign," lacking attention to the fundamental power dynamics and descriptions that affect the nexus and citizens' knowledge of it.

From this viewpoint, researchers approaching the FEW nexus emphasise the importance of recognising and discoursing difficulties in the learning process to allow stakeholders to interpret research results into concrete actions efficiently. The learning process perception intends to comprehend how public ideas towards the FEW nexus advance over time and how external effects form this understanding. While this perception has fewer publications, it holds significance in certifying the practical applications of the FEW nexus policies.

4.3 NEXUS GOVERNANCE APPLICATIONS WITHIN THE SOCIETY

Presently, research on nexus governance has mostly centred around the accessibility stage, mainly concentrating on connectivity issues (such as establishing relationships between elements) and innovation aspects (like growing enhanced solutions for urban or community problems). Remarkably, the primary assessment tools recognised include urban metabolism, footprints, and input/output analysis (Chini, Konar and Stillwell, 2017). Implementing these methods has resulted in innovative solutions, such as rooftop production, water-saving strategies, urban aquaponics, and floating production (Zanon et al., 2017). However, operationalising the idea of equitability has received limited consideration (Martínez-Guido, González-Campos and Ponce-Ortega, 2019). Table 4.1 provides the outline of nexus governance in the assessment stage.

Recently, attention in the awareness stage has been extended in three ways: participation, concentrating on stakeholder engagement (Agusdinata and Lukosch, 2019), and coordination, within the framework of connection across several sectors

TABLE 4.1
Outline of Nexus Governance Cases in the Assessment Stage

Principle	Description	Reference
Equitability	Politics, inequality, and power integration	Covarrubias, Spaargaren and Boas (2019)
	Justice and resilience	Schlör et al. (2018)
	Health and sustainability	Giatti et al. (2019)
	Access to resources in marginalised societies	Martínez-Guido, González-Campos and Ponce-Ortega (2019)
Innovation	Household water-saving strategy	Zhang et al., (2019)
	Rooftop production	Toboso-Chavero et al. (2019)
	Floating production	Zanon et al. (2017)
Connectivity	Input–output analysis	Feng et al. (2019)
	Urban metabolism	Zanon et al., (2017)
	Environmental footprint evaluation	Chini, Konar and Stillwell (2017)

and scales. Despite the challenges in linking nexus research with developmental and governance solutions, there is pressing attention to socially involved methods that share verdict progress and knowledge in the social and political framework (Schulterbrandt Gragg et al., 2018). Table 4.2 outlines cases of nexus governance in the awareness stage.

There is a rising demand to move the FEW nexus towards solutions-built applications at various stages. Accessibility is a severe field that shifts to nexus governance, indicating attention to policymaking, capacity, distributed management, technology, knowledge, authority, and resources (Elagib et al., 2021). While the urgency for accessibility is apparent, nexus governance may require incorporating legality, empowerment, and strategy layers that presently do not play a vital part in research (Taniguchi et al., 2019). Table 4.3 outlines cases of nexus governance in the accessibility stage.

A remarkable example is Roggema's work on design-led food-energy-water systems for urban locations (Roggema and Monti, 2021). The study shows innovative methods in different cities, such as an indigenous, purifying, reciprocal, hyperlocalised, and inclusive city system in western Sydney, a vegetable food supply system in

TABLE 4.2
Outline of Nexus Governance Cases in the Awareness Stage

Principle	Description	Reference
Coordination	Established a complete device for simplifying several sector discussions	Daher, Lee, et al. (2019)
Sharing	Ran field assessments to raise the outline of social challenges in urban systems	Ozturk (2015)
Participation	International workshops	Treemore-Spears et al. (2016)
	Role-playing game	Agusdinata and Lukosch (2019)

TABLE 4.3
Outline of Nexus Governance Cases in the Accessibility Stage

Principle	Description	Reference
Empowerment	Recognise performers who have the expertise and capacity for urban governance	Artioli, Acuto and McArthur (2017)
Strategy	Presented subsidies to incentivise irrigation	Taniguchi et al. (2019)
	Using a portable nexus to involve urban resources, technology, and knowledge and water-food-energy management	Roggema and Yan (2019)
Legitimacy	Regionalised management of water-food-energy resources	Lahmouri, Drewes and Gondhalekar (2019)
	Unified policymaking applicable to dangerous management and the water-food-energy nexus	Elagib et al. (2021)

Tokyo (Roggema and Monti, 2021), an altering food-energy-water system in Detroit (Thün, Sanyal and Velikov, 2021), a transformational forecasting method towards food-positive landscapes in Le Fouture de Groningen, and urban drifts in Rotterdam (Tillie and Roggema, 2021).

FEW nexus systems mostly show four features: (1) an extremely interrelated and interdependent multidimensional system of resources; (2) the integration of environmental, political, social, and economic dimensions; (3) the goal to maximise synergies and reduce trade-offs; and (4) the goal to have a more considerable effect on resource policy and management. Methods and solutions discoursing these features are mainly dependent on the specific case and influenced by political factors, considering the economy, local geography, climate, resource demand and supply.

4.4 BENEFITS OF FEW NEXUS GOVERNANCE IN THE SOCIETY

It is important to state that the governance theme is essential in solving complicated relationships within the FEW nexus in society. Despite the rising demands for more consideration of the policies and politics of nexus parts, researchers often overlook how trade-offs are transferred and decisions are established in practice, as illustrated in Figure 4.4. The nexus governance framework requires managing various policy connections, facilitating trade-offs, and encouraging synergies between resource management systems.

The society of science and practice is a platform for connectivity and discussion. It allows various stakeholders of education to consider several thematic sectors to improve on FEW nexus activities.

4.4.1 The Concept of Connectivity

Connectivity includes implementing a wide range of issues that cut across the limitations of natural resources. Natural resources are crucial in impacting or advancing society and the environment in various ways based on their capture, distribution,

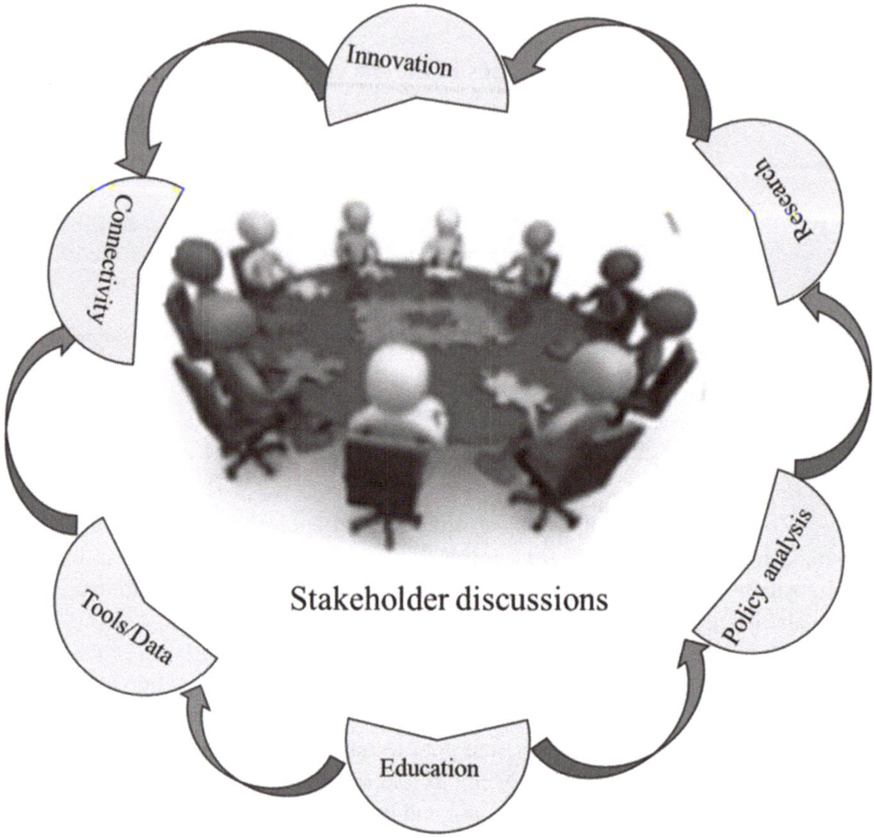

FIGURE 4.4 The nexus society of science and stakeholder discussion.

consumption, and disposal. The absence of consistency in broad policy commitment and direction brings about considerable losses or conflicts in natural resources and triggers unwanted trade-offs. As the drive of nexus governance creativities grows, the present research focuses on understanding and quantifying nexus connections [33]. Focusing on the premise that FEW systems are interrelated, nexus governance research attempts to reinforce food, energy, and water security by solidifying perceptions into nexus integration, cross-relationships, and adaptation (Al-Saidi and Elagib, 2017). Distinguished research by D'Odorico et al. (Soulé et al., 2004) identified critical relationships within the FEW nexus, such as reservoir operation, biofuels, and unusual fossil fuel withdrawal. In the same way, Yuan and Lo (2020) emphasised the importance of integrating these three items into a measurable metric to obtain an inadequacy path and bridge knowledge gaps in enhancing resource systems.

Several existing nexus research studies acknowledge the dominance of detecting the possible relationships and dependencies through data-informed evaluations and policy-making. Various research works apply quantitative modelling approaches to characterise the FEW nexus with growing complexity (Albrecht, Crootof and Scott, 2018).

These include methods like the input–output technique (Lin et al., 2019), life-cycle analysis (Yuan et al., 2018), GIS-based devices (Lin et al., 2019), the fusion of material flow and the life-cycle method (Chen, Feng and Ma, 2020), and system dynamics modelling. These methodologies mainly help the recognition of synergies and trade-offs within FEW systems, offering a basis for enhancing resource distribution and consumption by assessing serious cross-sector responses relating to food, energy, and water.

4.4.2 The Concept of Innovation

Innovation appears as the dynamic epicentres of nexus governance research. It indicates the action or procedure of paving the way for new and effective governance devices. The FEW nexus has increased technological innovation across different areas, encouraging the transmission of innovation and technology, building capacities for decision-makers, and institutionalising regional knowledge. The resultant innovation is an estimated result of integrating various views and devices into policymaking processes and outcomes (Apeh and Nwulu 2025). An initial assumption in nexus governance studies is that innovation contributes to harmonising the FEW nexus when it directly discourses waste elimination and resource circulation (da Rosa et al., 2021). These technologies intend to restrict water consumption while reclaiming substantial quantities of nutrients, increase crop advancement, under water stress conditions [46], and control communication and information technologies for more practical resource utilisation for human desires (Sarni, 2015). Different examples of technological innovation thrive: dry-cooling thermoelectric power plants, precision agriculture, renewable energy, economic innovations, irrigation management, water efficiency, recycling, and reuse technologies (Diaper et al., 2001). These innovations help sustainability in water purification and energy generation while minimising waste. While technological innovation often surrounds nexus governance studies, interdisciplinary methods are essential for simplifying technical issues and modelling methods applicable to these issues. A slight emphasis on engineering solutions shows that policymakers are insufficient in evaluating synergies and trade-offs between nexus connections. Coincidentally, both the public and private sectors assess this as a prospect for economic growth and business development, encouraging innovative collaborations to discourse nexus challenges jointly (e.g., the World Business Council for Sustainable Development, World Bank, and the World Economic Forum) (Schmidt and Matthews, 2018).

4.4.3 The Concept of Equitability

Equitability emphasises establishing equal opportunities to achieve economic, social, and ecological profits. While much interest has been directed towards the fairness principle regarding the distribution of different benefits, there is a notable lapse in considering benefit dispersal among investors in the decision-making process. A complete grip of this knowledge allows governance to sincerely adopt equal opportunities, bringing about equitable outcomes for society, the environment, and humans. Several research works emphasise the importance of women's voices and contributions in shaping suitable and sustainable resource organisations. For instance, men-owned enterprises utilise electricity more often and intensively than women. Realising the balanced sharing of men and women in energy involvement is important for poverty alleviation, sustainable growth, and well-being, and it involves an active contribution to decision-making processes (Pueyo, Carreras and Ngoo, 2020). Gendered functions meaningfully impact women in connection to water availability. Research highlights several types of women's empowerment that decrease the susceptibility of families to food uncertainty (Kevany and Huisingh, 2013). Pueyo et al. (2020) recognise multiple sources of gender inequity, such as access to resources, business characteristics, care responsibilities, motivations, and professional discrimination. Addressing these involves interdisciplinary and concerted methods. The importance of interrelated approaches, such as infrastructure investment, plan and legislation establishment, implementing gender equivalence and women's leadership, and developing planned involvements, is emphasised (Kevany and Huisingh, 2013).

4.4.4 The Concept of Participation

Participation refers to the complete and active involvement of pertinent rights holders and investors. Higher participation typically increases acceptance and public backup for secured regions. In nexus governance, participation is an essential issue as it demands the addition of appropriate stakeholders and rights holders, including businesses, local people, communities, NGOs, and youth. Furthermore, better governance and active participation indicate broad policymaking and adaptive management combined to restabilise power dynamics. Stakeholder involvement has increased importance in natural resource management research. Although definitions may differ, the procedure typically includes detecting, classifying, and discovering stakeholder relationships. Techniques such as inspecting legal authority lists and web pursuits help understand key stakeholders. Social network analysis categorises stakeholders and investigates their relationships (Daher, Hannibal, et al., 2019).

Many studies approach food-energy-water by collaborating with stakeholders. For example, Sušnik et al. (2018) applied a critical gaming technique to discover several nexuses with stakeholder participation, generating a harmless environment to assess the possible consequences of policy decisions. Bielicki et al. (2019) utilised stakeholder knowledge and contribution to moderate displeasure with the strategy growth process. Villamor et al. (2019) undertook research with scientists and modellers, reiterating the procedure with local stakeholders to integrate various forms of experience into the system algorithm. Yuan et al. (2018) conducted an FEW nexus sustainability in four cities, describing the critical part of renewable energy and technological innovation. These developments, co-designed, co-generated, and co-distributed, integrate stakeholders with various viewpoints.

4.4.5 The Concept of Coordination

Coordination comprises establishing relationships across various stages of actors involved in governance. In several cases, encouraging general good governance is connected to clear approaches and management processes that extend to numerous stages of actors. Moving from vertical governance concerns within the same ecosystem to horizontal attention relating to teamwork and agreement across various sectors is crucial to boosting the competence of strategy coordination devices. This indicates improved reliability and coordination of sectoral strategies.

In transboundary sectors where resource moves surpass national limits, one nation's governance activities and preservative policies can meaningfully influence other countries. Concerning the assumption that FEW are interrelated, nexus thinking drives FEW nexus research towards incorporating the governance systems of various areas. This incorporation includes improving cross-sectoral consistency and establishing coordination devices that harmonise plans among areas in national and sub-national governments (transportation, land, minerals, finance, agriculture, environmental protection, and communication). Such improvements allow nations to exploit synergies and reduce trade-offs in resource usage (Rasul, 2014). Definite methods comprise positioning cross-sectoral policies, joining incentive structures, and inspiring infrastructure expansion (Rasul, 2016). A crucial stage is to increase prevailing organisational structures to enable cooperation instead of creating fresh institutions or starting a new activity (Simpson and Jewitt, 2019).

4.4.6 The Concept of Sharing

Sharing involves information sharing, power sharing, and benefit sharing. Benefit sharing and power sharing revolve around distributing costs, duties, benefits, and rights within the authority and central government. Information distribution relates to how outcomes are spread in the media. Establishing an operative platform that encourages regional collaboration, information exchange, and knowledge sharing is vital for developing a nexus governance plan and application. Effective platforms are expanding globally, such as the Platform for Regional Integrated Modelling and Analysis (PRIMA) in the United States (Kraucunas et al., 2015), WEF Nexus Tool 2.0 in Qatar (Daher and Mohtar, 2018), GIS-based Regional Environmental Assessment Tool for Food-Energy-Water nexus (GREAT for FEW) in Taiwan, and the Nexus Solutions Tool (NEST) in India and Pakistan (Vinca et al., 2020)—regional centres and platforms centre on discovering different application paths at various levels. For instance, the Global Food and Water System (GFWS) platform permits the evaluation of food and water excesses and shortages (Grafton, Williams and Jiang, 2015). Countries are advised to be involved with regional experience hubs to evaluate, attract, and monitor food-energy-water challenges, navigating the right path.

4.4.7 The Concept of Legitimacy

Legitimacy denotes that the governing body possesses democratically or legally instructed authority. Lockwood et al. (2010) suggest that legitimacy includes three parts addressed by Bernstein (2004): the honesty with which this authority is exercised, the rationality of an organisation's expertise through stakeholder acceptance or democratic statute, and the decentralisation of power to the smallest and most efficiently exercised level. Legitimacy relies on an expert having reliable targets. These conditions of legitimacy highlight the means of an expert, increasing wide-ranging acceptance and gratitude in society. Governing agencies can attain legitimacy by partaking in processes from the grassroots, and they can improve their credibility by promoting community-based governance through consensus-building around a common idea. For laws to be regarded as legitimate, they must be recognised and enforced through legal procedures considered by responsible and authoritative management. Instead of entirely concentrating on governance ethics, assessing broader juridical and authorised measures is desirable as part of an appropriate agenda. For instance, a joint legislative agenda might involve complete regulations covering resource withdrawal, application, waste treatment discharge, and emissions (Olawuyi, 2020). This combination could streamline accrediting processes and simplify information distribution across several areas.

4.4.8 The Concept of Empowerment

A long-standing concern is evident in early governance and decision-making processes. Empowerment aims to simplify challenging governance choices by encouraging participatory and bottom-up approaches. Significantly, allocating more authority to local government systems is pivotal in the evolving governance paradigm (Withanachchi et al., 2018). Empowerment delineates and restricts the actors with participation or resource utilisation rights. As the empowerment literature identifies, accountability is key in enhancing and ensuring accomplishments within a governance system. May (2007) defines accountability structures for specific levels, encompassing legal, bureaucratic, professional, and political systems, aiming for consistency while promoting flexibility and innovation. Empowerment enables an exploration of the systematic responsibilities of different government levels, providing crucial insights for reflexive governance processes (Daher, Hannibal, et al., 2019).

4.4.9 The Concept of Strategy

The idea of the strategy revolves around a source to achieve long-standing decision-making aims. This involves technical knowledge, financial and managerial skills, resources, succession planning, leadership, management systems, and training availability that facilitate mechanisms that align with purposes (Lockwood et al., 2010). It targets developing inducements and offers a source to stimulate actors to compare knowledge, meditate, study, and adapt their behaviour over time.

Strategy has gained broad support and recognition in nexus governance, enabling adaptation to changing internal or external conditions. It allows a positive response to change, decreasing unpredicted challenges. Moreover, it supports internal systematic self-reflection, evaluating the dependability of assumptions, making informed decisions, and accomplishing improved results (Lockwood et al., 2010).

4.5 THE SIGNIFICANCE OF FEW NEXUS IN URBAN–RURAL SYSTEMS

Although urban and rural areas are geographically distributed, their complex relationship describes their workability. They benefit from multiple resources, which are typically balanced and interconnected through a wide range of relationships.

Urban–rural relationships have gained extensive recognition in global policy agendas, such as the SDGs (Baffoe et al., 2021) and the New Urban Agenda (UN-Habitat, 2017). As the SDGs search for complementary regional development, urban–rural relationships are vital in modern socio-environmental variations. This importance goes beyond attaining SDG 11 (Sustainable Cities and Communities) to impacting SDG 2 (Zero Hunger) and SDG 12 (Responsible Consumption and Production) (UN-Habitat, 2017).

The exponential increase in goods, lifestyles, people, capital, ideas, services, and technology between urban and rural areas shows their spatial increase and regional linkage. Food, energy, and water demand gradually grow in urban areas, pushing these natural systems towards critical limits. This phenomenon weakens the resilience of both urban and rural systems, thereby exposing the inhabitants to the environmental disaster.

Figure 4.5 illustrates the theoretical knowledge of the FEW nexus in the framework of urban–rural systems combined with evolving pressures and related differences.

The study conceives that due to communal environmental resources, urban–rural systems are collectively susceptible to the consequences of environmental degradation, though at varying levels. The study by Wilbanks et al. (2013) likewise delves into how disaster influences in distant asset regions can adversely affect dependent urban areas and vice versa. As the urban and rural systems are gradually inter-reliant, shared resilience, connected to interactive bonds and networks, has become essential in regional development planning. In reaction to the increasing urban population and fluctuating social dynamics, there is a growing necessity to control the local reserve of natural resources. Otherwise, urban–rural systems may end up contending with each other, escalating the combinatorial difficulty of the FEW nexus. The contradictory

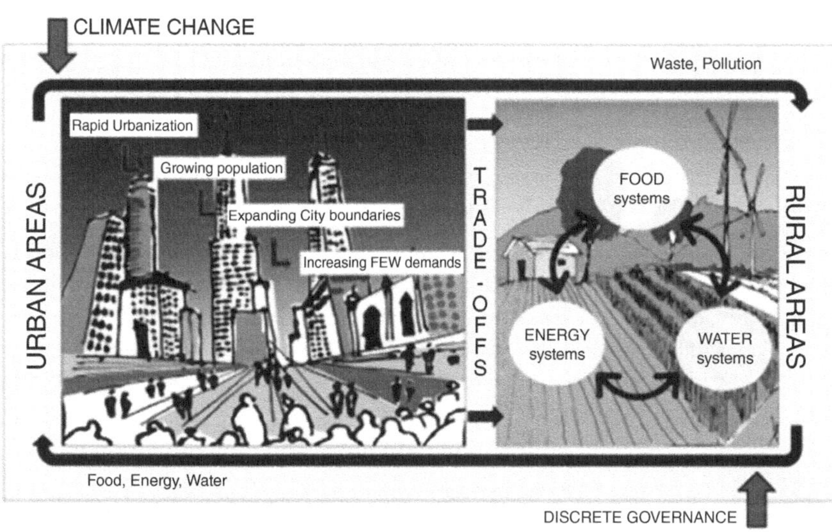

FIGURE 4.5 Theoretical knowledge of the FEW nexus in the framework of urban-rural systems.

method of urban–rural expansion and the sectoral tactic to the FEW nexus (Pittock, Hussey and McGlennon, 2013) have brought about discrepancies.

Consequently, rural systems are currently unduly influenced by climate change. Confirming a safe and constant supply of food, energy, and water for both urban and rural areas while sustaining a lively environment is, at present, the fundamental challenge for regional authorities. Similarly, there is an increasing requirement for combined, nexus-aware systematic contexts to better measure and model the connections between FEW sectors at the regional level. However, prevailing agendas seemingly fall short of resolving the spatio-temporal scales of asset management. Besides, the study suggests that enhancing FEW resources in urban and rural areas basically relies on the effectiveness of FEW nexus thinking at the base level.

4.6 STRATEGIES FOR ENHANCING SOCIETAL BENEFITS IN THE FEW NEXUS

The FEW nexus represents the interrelationship of these critical resources, stressing the necessity for integrated methods to tackle challenges and exploit benefits for society. As populations increase and environmental threats increase, developing plans that increase societal benefits within this nexus becomes important. This section presents different strategies intended to reach the purpose.

4.6.1 Integrated Resource Management

Integrated resource management includes deliberating on the interrelationships between food, energy, and water systems to optimise resource application and reduce adverse environmental effects. For example, implementing activities including precision agriculture can minimise water utilisation and improve crop production through designed usages of resources such as water and fertilisers. In the same way, renewable energy applications can assist in alleviating environmental dilapidation related to traditional energy sources while decreasing water usage in energy production procedures (Apeh, Meyer and Overen, 2021).

4.6.2 Innovation and Technology

Innovation and technology are vital in improving societal benefits within the FEW nexus by determining efficiency improvements and enabling new solutions. For example, advancements in agricultural biotechnology have led to the development of drought-resistant crops that can thrive in water-stressed environments, thus enhancing food security. Similarly, solar panels and wind turbines have revolutionised energy production, providing clean and renewable alternatives to fossil fuels.

4.6.3 Cross-sector Collaboration

Cross-sector collaboration comprises partnerships among stakeholders, such as academia, governments, civil society, and businesses, to tackle challenges within the FEW nexus. Collaboration can result in broad solutions that aim for several perceptions and influence various skills and resources. For example, a collaborative effort between water utilities, energy companies, and agricultural producers could initiate the expansion of integrated systems that employ agricultural waste to generate bioenergy and recycle wastewater for irrigation.

4.6.4 Community Engagement and Empowerment

Engaging and empowering communities within the FEW nexus is important for assuring that plans are tailored to local requirements and priorities. This can include raising awareness regarding resource challenges through education and outreach plans, offering training possibilities for sustainable performances, and backing up community-led creativities, including community gardens or renewable energy companies. By including communities in decision-making processes, approaches can be more comprehensive and efficient in determining local challenges.

4.6.5 Policy and Governance Transformation

Policy and governance agendas are critical in determining results within the FEW nexus. Transformations intended at encouraging incorporated arrangements, incentivising sustainable practices, and tackling systemic inequities can improve societal benefits. This may include executing strategies such as water pricing devices that deliberate the actual price of resource usage, creating regulatory standards for environmental protection, and fostering adaptive governance structures that can respond to changing situations and uncertainties.

4.6.6 Resilience Building

Building resilience within the FEW nexus comprises preparing for and acclimatising to interruptions such as natural disasters, climate change, and financial shocks. This can comprise expanding resource collections to decreasing reliance on single sources, investing in infrastructure that can tolerate severe events, for example, floods or droughts, and encouraging adaptive management methods that allow flexible responses to changing conditions. Furthermore, social security nets and insurance devices can assist communities in managing negative effects and improving more rapidly from disasters, guaranteeing that societal benefits are maintained despite challenges.

4.7 CONCLUSION

The search for societal-inclusive sustainability through interdisciplinary approaches in FEW nexus research and practice is vital for tackling the intricate challenges of our time. By strengthening collaboration among various stakeholders, sharing knowledge from different disciplines, and prioritising the desires of societies, it is possible to forge a pathway towards a more equitable, resilient, and environmentally sustainable future. Implementing the interrelation of FEW systems and recognising their deep influence on society suggests possibilities to create innovative solutions that encourage both human well-being and ecological health. As society progresses, it is crucial to continue advancing interdisciplinary methods, involving local communities and supporting policies that support societal-inclusive sustainability. A more affluent and harmonious world could be built for present and future generations through collective effort and integrated commitment. However, sustainable development is directly connected to one of the most modern challenges in academic research and for society: the rational consumption of natural resources. Through the principles incorporated within, the FEW nexus strategies can act as a lead and contribute to the advancement of sustainable development. Improving societal benefits within the FEW nexus needs an inclusive and combined method that addresses the intricate connections between FEW systems. By implementing policies such as innovation, integrated resource management, technology implementation, cross-sector collaboration, community participation, strategy transformation, and resilience building, societies can attain more sustainable and reasonable results that meet the requirements of current and prospective generations.

REFERENCES

Agusdinata, D. B. and Lukosch, H. (2019) 'Supporting interventions to reduce household. greenhouse gas emissions: A transdisciplinary role-playing game development'. *Simulation & Gaming*, 50(3), pp. 359–376.

Albrecht, T. R., Crootof, A. and Scott, C. A. (2018) 'The water-energy-food nexus: A systematic review of methods for nexus assessment'. *Environmental Research Letters*, 13(4), p. 43002.

Alfieri, L. et al. (2018) 'Multi-model projections of river flood risk in Europe under global warming'. *Climate*, 6(1), p. 6.

Al-Saidi, M. and Elagib, N. A. (2017) 'Towards understanding the integrative approach of the water, energy and food nexus'. *Science of the Total Environment*, 574, pp. 1131–1139.

Apeh, O. O., Meyer, E. L. and Overen, O. K. (2021) 'Modeling and experimental analysis of battery charge controllers for comparing three off-grid photovoltaic power plants'. *Heliyon*, 7(11), e08331.

Apeh, O. O., & Nwulu, N. I. (2024). The water-energy-food-ecosystem nexus scenario in Africa: Perspective and policy implementations. *Energy Reports*, 11, 5947–5962.

Apeh, O. O., & Nwulu, N. I. (2025). Improving traceability and sustainability in the agri-food industry through blockchain technology: A bibliometric approach, benefits and challenges. *Energy Nexus*, 17, 100388.

Artioli, F., Acuto, M. and McArthur, J. (2017) 'The water-energy-food nexus: An integration agenda and implications for urban governance'. *Political Geography*, 61, pp. 215–223.

Baffoe, G. et al. (2021) *Urban-rural linkages: Effective solutions for achieving sustainable development in Ghana from an SDG interlinkage perspective*. Springer.

Barton, B. (2015) *Feeding ourselves thirsty: How the food sector manages global water risks*. Ceres.

Bernstein, S. (2004) 'Legitimacy in global environmental governance'. *Journal of International Law & International Relations*, 1, p. 139.

Bielicki, J. M. et al. (2019) 'Stakeholder perspectives on sustainability in the food-energy-water nexus'. *Frontiers in Environmental Science*, 7, p. 7.

Cairns, R. and Krzywoszynska, A. (2016) 'Anatomy of a buzzword: The emergence of "the water-energy-food nexus" in UK natural resource debates'. *Environmental Science & Policy*, 64, pp. 164–170.

Chen, C.-F., Feng, K.-L. and Ma, H. (2020) 'Uncover the interdependent environmental. impacts associated with the water-energy-food nexus under resource management strategies'. *Resources, Conservation and Recycling*, 160, p. 104909.

Chen, H. and Carter, K. E. (2016) 'Water usage for natural gas production through hydraulic fracturing in the United States from 2008 to 2014'. *Journal of Environmental Management*, 170, pp. 152–159.

Chini, C. M., Konar, M. and Stillwell, A. S. (2017) 'Direct and indirect urban water footprints of the United States'. *Water Resources Research*, 53(1), pp. 316–327.

Covarrubias, M., Spaargaren, G. and Boas, I. (2019) 'Network governance and the urban the nexus of water, energy, and food are lessons from Amsterdam'. *Energy, Sustainability and Society*, 9(1), pp. 1–11.

da Rosa, F. S. et al. (2021) 'Environmental innovation and the food, energy and water nexus in the food service industry'. *Resources, Conservation and Recycling*, 166, p. 105350.

Daher, B., Hannibal, B., et al. (2019) 'Toward creating an environment of cooperation between water, energy, and food stakeholders in San Antonio'. *Science of the Total Environment*, 651, pp. 2913–2926.

Daher, B., Lee, S.-H., et al. (2019) 'Towards bridging the water gap in Texas: A water-energy-food nexus approach'. *Science of the Total Environment*, 647, pp. 449–463.

Daher, B. T. and Mohtar, R. H. (2018) 'Water–energy–food (WEF) nexus tool 2.0: Guiding integrative resource planning and decision-making'. In Anik Bhaduri, Claudia Ringler, Ines Dombrowsky, Rabi Mohtar, and Waltina Scheumann (Eds.), *Sustainability in the water energy food nexus* (pp. 36–59). Routledge.

De Strasser, L. et al. (2016) 'A methodology to assess the water energy food ecosystems nexus in transboundary river basins'. *Water*, 8(2), p. 59.

Del Borghi, A., Moreschi, L. and Gallo, M. (2020) 'Circular economy approach to reduce water–energy–food nexus'. *Current Opinion in Environmental Science & Health*, 13, pp. 23–28.

DeNicola, E. et al. (2015) 'Climate change and water scarcity: The case of Saudi Arabia'. *Annals of Global Health*, 81(3), pp. 342–353.

Diaper, C. et al. (2001) 'Water-recycling technologies in the UK'. *Water Environment Journal*, 15(4), pp. 282–286.

Elagib, N. A. et al. (2021) 'Exploring the urban water-energy-food nexus under environmental hazards within the Nile'. *Stochastic Environmental Research and Risk Assessment*, 35, pp. 21–41.

FAO, IFAD, and WFP. (2015). *The State of Food Insecurity in the World 2015: Meeting the 2015 International Hunger Targets: Taking Stock of Uneven Progress*. Rome: FAO.

Fasel, M. et al. (2016) 'Blue water scarcity in the Black Sea catchment: Identifying key actors in the water-ecosystem-energy-food nexus'. *Environmental Science & Policy*, 66, pp. 140–150.

Feng, C. et al. (2019) 'Uncovering urban food-energy-water nexus based on physical input-output analysis: The case of the Detroit Metropolitan Area'. *Applied Energy*, 252, p. 113422.

Foran, T. (2015) 'Node and regime: Interdisciplinary analysis of water-energy-food nexus in the Mekong region'. *Water Alternatives*, 8(1), pp. 655–674.

Geissdoerfer, M. et al. (2017) 'The circular economy – A new sustainability paradigm?' *Journal of Cleaner Production*, 143, pp. 757–768.

Ghodsvali, M., Krishnamurthy, S. and de Vries, B. (2019) 'Review of transdisciplinary approaches to food-water-energy nexus: A guide towards sustainable development'. *Environmental Science & Policy*, 101, pp. 266–278.

Giampietro, M., Aspinall, R. J., Bukkens, S. G. F., Cadillo Benalcazar, J. J., Díaz Maurin, F., Flammini, A., Gomiero, T., Kovacic, Z., Madrid-Lopez, C., Ramos Martin, J., & Serrano Tovar, E. T. (2013). *An innovative accounting framework for the food-energy-water nexus: Application of the MuSIASEM approach to three case studies* (Environment and Natural Resources Management Working Paper No. 56). Food and Agriculture Organization of the United Nations. https://www.fao.org/3/i3468e/i3468e.pdf

Giatti, L. L. et al. (2019) 'Nexus of exclusion and challenges for sustainability and health in an urban periphery in Brazil'. *Cadernos De Saude Publica*, 35, e00007918.

Grafton, R. Q., Williams, J. and Jiang, Q. (2015) 'Food and water gaps to 2050: Preliminary results from the global Food and water system (GFWS) platform'. *Food Security*, 7, pp. 209–220.

Gustavsson, J. et al. (2011) *Global food losses and food waste*. FAO Rome.

Halbe, J. et al. (2018) 'Governance of transitions towards sustainable development–the water–energy–food nexus in Cyprus'. In R. H. Mohtar, M. A. Hanjra, and J. Hoekstra (Eds.), *Sustainability in the water energy food nexus* (pp. 165–182). Routledge.

Hannibal, B. and Vedlitz, A. (2018) 'Throwing it out: Introducing a nexus perspective in examining citizen perceptions of organizational food waste in the US'. *Environmental Science & Policy*, 88, pp. 63–71.

Hassanpour Adeh, E., Selker, J. S. and Higgins, C. W. (2018) 'Remarkable agrivoltaic influence on soil moisture, micrometeorology and water-use efficiency'. *PLoS One*, 13(11), p. e0203256.

Hawksworth, J., & Chan, D. (2015). *The World in 2050: Will the shift in global economic power continue?* PwC Economics and Policy Team. Retrieved from https://www.pwc.com/ee/et/publications/pub/world-in-2050-february-2015.pdf

Healy, R. W. et al. (2015) 'The water-energy nexus: An earth science perspective'. In *US geological survey circular* (p. 1407). US Department of the Interior, US Geological Survey.

Heidari, A. and Khovalyg, D. (2020) 'Exploring water-energy nexus at the building level'. *ASHRAE Transactions*, 126(2), pp. 308–314.

Henry, A. D. (2009) 'The challenge of learning for sustainability: A prolegomenon to theory'. *Human Ecology Review*, 16(2), pp. 308–314.

Kevany, K. and Huisingh, D. (2013) 'A review of progress in the empowerment of women in rural water management decision-making processes'. *Journal of Cleaner Production*, 60, pp. 53–64.

Kraucunas, I. et al. (2015) 'Investigating the nexus of climate, energy, water, and land at decision-relevant scales: The platform for regional integrated modeling and analysis (PRIMA)'. *Climatic Change*, 129, pp. 573–588.

Kummu, M. et al. (2012) 'Lost food, wasted resources: Global food supply chain losses and their impacts on freshwater, cropland, and fertiliser use'. *Science of the Total Environment*, 438, pp. 477–489.

Lahmouri, M., Drewes, J. E. and Gondhalekar, D. (2019) 'Analysis of greenhouse gas emissions in centralized and decentralized water reclamation with resource recovery strategies in Leh Town, Ladakh, India, and potential for their reduction in context of the water–energy–food nexus'. *Water*, 11(5), p. 906.

Lin, Y.-C. et al. (2019) 'Comprehensive assessment of regional food-energy-water nexus with GIS-based tools'. *Resources, Conservation and Recycling*, 151, p. 104457.

Lockwood, M. et al. (2010) 'Governance principles for natural resource management'. *Society and Natural Resources*, 23(10), pp. 986–1001.

Mannan, M. et al. (2018) 'Quantifying the energy, water and food nexus: A review of the latest Developments based on life-cycle assessment'. *Journal of Cleaner Production*, 193, pp. 300–314.

Martínez-Guido, S. I., González-Campos, J. B. and Ponce-Ortega, J. M. (2019) 'Strategic planning to improve the human development index in disenfranchised communities through satisfying food, water, and energy needs'. *Food and Bioproducts Processing*, 117, pp. 14–29.

Mathioudakis, V. et al. (2017) 'The water footprint of second-generation bioenergy: A comparison of biomass feedstocks and conversion techniques'. *Journal of Cleaner Production*, 148, pp. 571–582.

May, P. J. (2007) 'Regulatory regimes and accountability'. *Regulation & Governance*, 1(1), pp. 8–26.

Middleton, C. et al. (2015) 'The rise and implications of the water-energy-food nexus in Southeast Asia through an environmental justice lens'. *Water Alternatives*, 8(1).

Olawuyi, D. (2020) 'Sustainable development and the water-energy-food nexus: Legal challenges and emerging solutions'. *Environmental Science & Policy*, 103, pp. 1–9.

Ozturk, I. (2015) 'Sustainability in the food-energy-water nexus: Evidence from BRICS (Brazil, the Russian Federation, India, China, and South Africa) countries'. *Energy*, 93, pp. 999–1010.

Pena-Arancibia, J. L. et al. (2020) 'Groundwater use and rapid irrigation expansion in a changing climate: Hydrological drivers in one of the world's food bowls'. *Journal of Hydrology*, 581, p. 124300.

Pittock, J., Hussey, K. and McGlennon, S. (2013) 'Australian climate, energy and water policies: Conflicts and synergies'. *Australian Geographer*, 44(1), pp. 3–22.

Portney, K. E. et al. (2018) 'Awareness of the food–energy–water nexus and public policy support in the United States: Public attitudes among the American people'. *Environment and Behavior*, 50(4), pp. 375–400.

Pueyo, A., Carreras, M. and Ngoo, G. (2020) 'Exploring the linkages between energy, gender, and enterprise: Evidence from Tanzania'. *World Development*, 128, p. 104840.

Rasul, G. (2014) 'Food, water, and energy security in South Asia: A nexus perspective from the Hindu Kush Himalayan region☆'. *Environmental Science & Policy*, 39, pp. 35–48.

Rasul, G. (2016) 'Managing the food, water, and energy nexus for achieving the sustainable development goals in South Asia'. *Environment and Development*, 18, pp. 14–25.

Raworth, K. (2017) *Doughnut economics: Seven ways to think like a 21st-century economist*. Chelsea Green Publishing.

Rodriguez, D. J. et al. (2013) *Thirsty energy*. World Bank.

Rodríguez, L. F. et al. (2019) 'The development of the INFEWS-ER: A virtual resource center for transdisciplinary graduate student training at the nexus of food, energy, and water'. *Frontiers in Environmental Science*, 7, p. 38.

Roggema, R. and Monti, S. (2021) 'Nature driven planning for the FEW-nexus in Western Sydney'. In Rob Roggema (Ed.), *TransFEWmation: Towards design-led food-energy-water systems for future urbanization* (pp. 59–94). Springer.

Roggema, R. and Yan, W. (2019) 'Developing a design-led approach for the food-energy-water nexus in cities'. *Urban Planning*, 4(1), pp. 123–138.

Saladini, F. et al. (2018) 'Linking the water-energy-food nexus and sustainable development indicators for the Mediterranean region'. *Ecological Indicators*, 91, pp. 689–697.

Sarni, W. (2015). *Deflecting the scarcity trajectory: Innovation at the water, energy, and food nexus. Deloitte Review*, Issue 17. Available at: https://www2.deloitte.com/content/dam/insights/us/articles/water-energy-food-nexus/DUP1205_DR17_DeflectingtheScarcityTrajectory.pdf

Schlör, H., Venghaus, S. and Hake, J.-F. (2018) 'The FEW-nexus city index–Measuring urban resilience'. *Applied Energy*, 210, pp. 382–392.

Schmidt, J. J. and Matthews, N. (2018) 'From state to system: Financialization and the water-energy-food-climate nexus'. *Geoforum*, 91, pp. 151–159.

Schramski, J. R., Woodson, C. B. and Brown, J. H. (2020) 'Energy use and the sustainability of intensifying food production'. *Nature Sustainability*, 3(4), pp. 257–259.

Schulterbrandt Gragg, R. et al. (2018) 'A conceptualization of the urban food-energy-water nexus sustainability paradigm: Modeling from theory to practice'. *Frontiers in Environmental Science*, 6, p. 133.

Simpson, G. B. and Jewitt, G. P. W. (2019) 'The water-energy-food nexus in the anthropocene: Moving from 'nexus thinking to "nexus action"'. *Current Opinion in Environmental Sustainability*, 40, pp. 117–123.

Soulé, M. E. et al. (2004) 'The role of connectivity in Australian conservation'. *Pacific Conservation Biology*, 10(4), pp. 266–279.

Sušnik, J. et al. (2018) 'Multi-stakeholder development of a serious game to explore the water-Energy-food-land-climate nexus: The SIM4NEXUS approach'. *Water*, 10(2), p. 139.

Tabatabaie, S. M. H., Bolte, J. P. and Murthy, G. S. (2018) 'A regional scale modeling framework combining biogeochemical model with life cycle and economic analysis for integrated assessment of cropping systems'. *Science of the Total Environment*, 625, pp. 428–439.

Taniguchi, M. et al. (2019) 'Recovery of lost nexus synergy via payment for environmental services in Kumamoto, Japan'. *Frontiers in Environmental Science*, 7, p. 28.

Thün, G., Sanyal, T. and Velikov, K. (2021) 'Mapping the FEW-nexus across cascading scales: Contexts for Detroit from region to city'. In Rob Roggema (Ed.), *TransFEWmation: Towards design-led food-energy-water systems for future urbanization* (pp. 171–207). Springer.

Tillie, N. and Roggema, R. (2021) 'Synergetic planning and designing with urban FEW-flows: Lessons from Rotterdam'. In Rob Roggema (Ed.), *TransFEWmation: Towards design-led food-energy-water systems for future urbanization* (pp. 125–144). Springer.

Toboso-Chavero, S. et al. (2019) 'Towards productive cities: Environmental assessment of the food-energy-water Nexus of the urban roof mosaic'. *Journal of Industrial Ecology*, 23(4), pp. 767–780.

Treemore-Spears, L. J. et al. (2016) 'A workshop on transitioning cities at the food-energy-water nexus'. *Journal of Environmental Studies and Sciences*, 6, pp. 90–103.

UN ESCAP. (2017) *World economic situation and prospects 2017*. United Nations.

UN-Habitat (2017) *Implementing the new urban agenda by strengthening urban-rural linkages – Leave no one and no space behind* (88 pp). UN-Habitat.

United Nations (2015) *Reconciling resource uses in transboundary basins: Assessment of the water-food-energy- ecosystems nexus in the Sava River Basin* (pp. 1–121). UNECE. http://www.unece.org/fileadmin/DAM/env/water/publications/WAT_Nexus/ece_mp.wat_46_eng.pdf

United Nations General Assembly. (2015). *Transforming our world: The 2030 Agenda for Sustainable Development*. Resolution A/RES/70/1, adopted on 21 October 2015. United Nations.

Venghaus, S. and Hake, J.-F. (2018) 'Nexus thinking in current EU policies – The interdependencies among food, energy and water resources'. *Environmental Science & Policy*, 90, pp. 183–192.

Villamor, G. B. et al. (2019) *Contrasting stakeholder and scientist conceptual models of food-energy-water systems: A case study in Magic Valley, Southern Idaho* (p. 2). Socio-Environmental Systems Modelling.

Vinca, A. et al. (2020) 'The NExus Solutions Tool (NEST) v1. 0: An open platform for optimizing multi-scale energy–water–land system transformations'. *Geoscientific Model Development*, 13(3), pp. 1095–1121.

Walsh, B. P., Murray, S. N. and O'Sullivan, D. T. J. (2015) 'The water-energy nexus, an ISO50001 water case study and the need for a water value system'. *Water Resources and Industry*, 10, pp. 15–28.

Walton, M. (2018) *Energy has a role to play in achieving universal access to clean water and sanitation*. International Energy Agency (IEA).

Weitz, N. et al. (2017) 'Closing the governance gaps in the water-energy-food nexus: Insights from integrative governance'. *Global Environmental Change*, 45, pp. 165–173.

White, D. D. et al. (2017) 'Stakeholder analysis for the food-energy-water nexus in Phoenix, Arizona: Implications for nexus governance'. *Sustainability*, 9(12), p. 2204.

Wilbanks, T. et al. (2013) 'Climate change and infrastructure, urban systems'. In Thomas J. Wilbanks and Steven Fernandez (Eds.), *And vulnerabilities: Technical report for the US department of energy in support of the national climate assessment*. Springer.

Willis, H. H. et al. (2016) *Developing the Pardee RAND food-energy-water security index: Toward a global standardized, quantitative, and transparent resource assessment*. RAND.

Withanachchi, S. S. et al. (2018) 'A paradigm shift in water quality governance in a transitional context: A critical study about the empowerment of local governance in Georgia'. *Water*, 10(2), p. 98.

Yang, J. et al. (2018) 'Quantifying the sustainability of water availability for the water-food-energy-ecosystem nexus in the Niger River Basin'. *Earth's Future*, 6(9), pp. 1292–1310.

Yuan, K.-Y. et al. (2018) 'Spatial optimization of the food, energy, and water nexus: A life cycle assessment-based approach'. *Energy Policy*, 119, pp. 502–514.

Yuan, M.-H. and Lo, S.-L. (2020) 'Developing indicators for the monitoring of the sustainability of food, energy, and water'. *Renewable and Sustainable Energy Reviews*, 119, p. 109565.

Zanon, B. D. B. et al. (2017) 'Potential of floating production for delta and coastal cities'. *Journal of Cleaner Production*, 151, pp. 10–20.

Zhang, P. et al. (2019) 'Food-energy-water (FEW) nexus for urban sustainability: A comprehensive review'. *Resources, Conservation and Recycling*, 142, pp. 215–224.

5 Optimising the Food-Energy-Water Nexus

A Systematic Literature Review of Sustainable Development Perceptions and Global Disparities

5.1 INTRODUCTION

Globally, the demand for FEW is rising exponentially by about 50% due to urbanisation, differences in consumption style, climate change, and population growth (Vahabzadeh, Afshar and Molajou, 2023). In 2019, the UN recorded that approximately 821 million people in 2017 were underfed compared to 784 million inhabitants in 2015. Unfortunately, supplying food to these undernourished people worldwide will require more land than currently available (Uhlenbrook and Connor, 2019). Similarly, water accessibility is still a major concern in the global community, as the statistical reports show that in 2019, 785 million people lacked safe drinking water, while about two billion people had no water supply. Also, the UN projected that about 700 million people may be forced to migrate or seek refuge elsewhere due to severe water scarcity (Desa, 2018). From a more critical perspective, the World Bank reported that in 2019, about 840 million rural dwellers lacked electricity supply (Al-Rousan, Isa and Desa, 2018), while around three billion people had no means to refine cooking fuels, including renewable energy (Apeh, Meyer and Overen, 2022). Studies show that the Earth's surface temperature in 2018 was around 1°C, which is higher than the estimated measure by the Paris Agreement (Agbor et al., 2023). Hence, if this situation persists, it is projected that Arctic glaciers will melt, leading to an increase in sea and ocean levels and the submergence of 150 million people by 2050 and 360 million by 2100 (Kulp and Strauss, 2019).

Regarding this situation, the demand for FEW has started gaining popularity because the rising socio-economic pressures have surpassed the environment's capacity to recover, creating terrible costs for SDGs. To this end, research on optimising environmental resources is gaining worldwide attention as a means to mitigate threats such as waste, depletion of resources, and emissions. Previously, researchers have learnt the complicated corroboration of the FEW in the environment, as presented in Figure 5.1. This knowledge was piloted by nexus (Helmstedt et al., 2018). The term

DOI: 10.1201/9781003633846-5

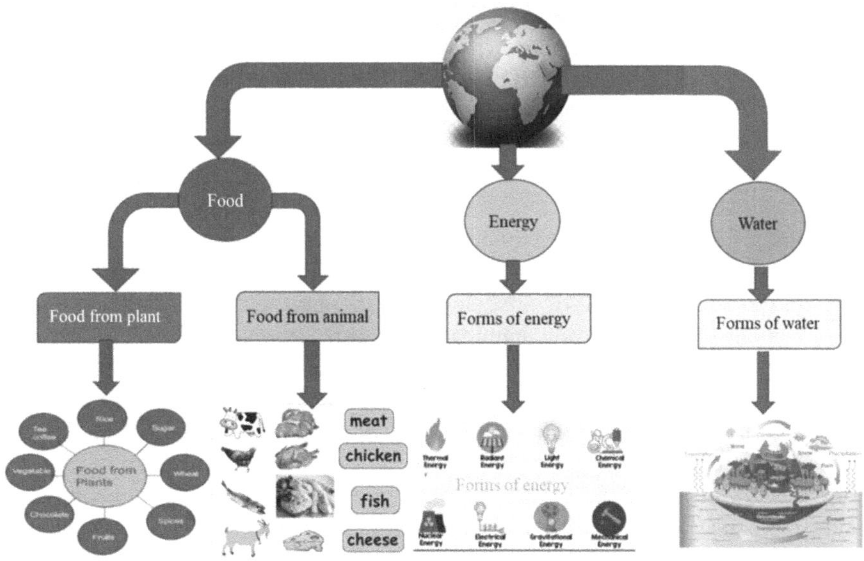

FIGURE 5.1 Description of food, energy, and water structure.

"nexus" was initiated as a requisite to serve a complete and integrated approach to the entire system rather than its elements to advance a viable result for the prospect of society (Albrecht, Crootof and Scott, 2018). Therefore, the idea of the FEW nexus has become a focus of interest in academic literature, workshops, policy-making platforms, and conferences as a better way to comprehend multipart correlations among numerous resource systems.

5.1.1 The Context of Nexus

The rate at which water is extracted and the strength of energy utilisation determine the procedures for food production. Just like food and water are the requirements for generating energy, it is the same way effective food and energy applications are required for water security. Meanwhile, water is powered by electricity generation, such as hydropower, fossil fuel extractions, purifications, and the nurturing of feedstock as applied to biofuel generation. Birol and Das (2010) stated that almost 15% of international water extractions are used for energy generation (2010). Similarly, energy is vital for extracting, supplying, treating, and transferring consumed and unconsumed water for societal utilisation. On the other hand, in the global community, approximately 8% of energy is used during water drives (Heilig, 2012). Water is essential for food generation and refining (crop irrigation, pond filling for aquaculture generation, and food refining); it relates to food generation and river systems by excess conveying sediments, agrochemicals, saline trading, and organic trading matter (Moss et al., 2008).

Aquastat (2011) projected that approximately 70% of international water extractions enter food generation. For instance, the decline in the quality of water is described as due to the rise in seafood generation (aquaculture) and pollution of water as a result of nutrient pileup through releases of unused fish food and fish misuse in water bodies (Aquastat, 2011; Troell et al., 2014; Peng, 2017). Similarly, the energy required by the food generation systems, including food generation and transportation, accounts for approximately 30% of the international energy generated (Ajanovic, 2011). Figure 5.2 demonstrates the relationship between the FEW nexus.

Similarly, food can be utilised for energy generation, as in the case of biofuel, and a study shows that approximately 1% of food supplied is utilised for a particular objective (Garcia and You, 2016). Additionally, these present undesirable outcomes of FEW applications, such as contamination of water systems and the discharge of GHGs (Chhipi-Shrestha, Hewage and Sadiq, 2017). However, reactions from climate change, such as temperature increase, differences in the pattern of rainfall, and extreme weather conditions, influence FEW generations (Apeh and Nwulu, 2025). For instance, an increase in temperature causes air to warm up, which retains more moisture and can increase the chances of rainfall, causing flooding and reducing food yield (IPCC and Team CW, 2014). On the other hand, differences in climate, such as warmer weather conditions, cause a high rate of evaporation and decreased groundwater and surface water levels, producing more droughts and water scarcity. Consequently, animals face direct threats, along with the spread of parasites and diseases associated with the fluctuations in weather conditions, which can greatly

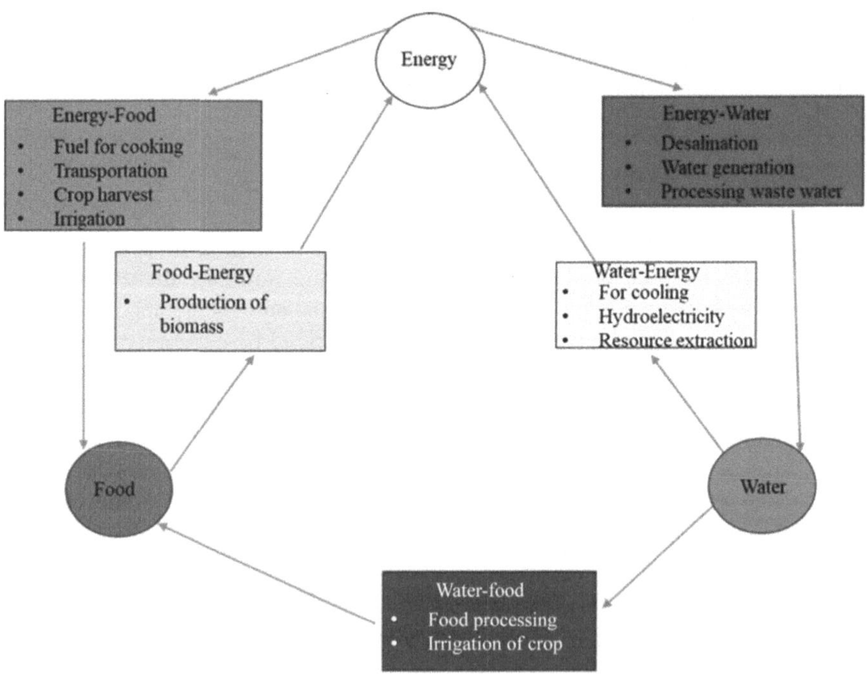

FIGURE 5.2 The connections among the domain of the FEW nexus.

affect food yields (Arthur et al., 2019). Equally, the decreasing water system due to excessive climate conditions influences hydro-energy generation and biomass utilised in biofuel generation (IPCC and Team CW, 2014).

5.1.2 The Overview of Food-Energy-Water Nexus Advancement

There have been various perspectives on FEW proposals for over a decade now, with noticeable applications in different scientific fields, including water for food and the nexus of political and socialist civilisations (Wittfogel, 1955; Lele, Klousia-Marquis and Goswami, 2013). The scientific field has emphasised the interrelationship among FEW since the oil disaster in the 1970s. The research study by Meadows et al. (1972) used a simulation method on the sudden increase in the population and financial systems with limited water and food, and the results showed that a constant increase in the economy would bring about a serious shortage of resources (Meadows et al., 1972). Afterwards, another research by Levinson et al. (1977) deliberated on the circumstances leading to energy pressure and water generation in three developing nations (Upper Volta, Mali, and Nepal) (Levinson, Rosenberg and Yansane, 1977). As the three-point FEW nexus evolves, the researchers steadily neglected separate analytical methods and implemented systematic approaches, creating the study viewpoint for successive FEW nexus activities (David et al., 2022).

However, from 1980 to 2010, there was a noticeable gradual research interest and project growth on the FEW nexus globally, including nations such as India, Brazil, the United States, and Japan. Between 1983 and 1986, a sequence of strategies regarding the nexus of energy was instituted by the UNU to regulate their connection, such as the food–energy relationships (Sachs and Silk, 1990). Concurrently, some international establishments tried to deliberate the FEW nexus's social, economic, environmental, and political frameworks. For example, in 1986, the next global conference, titled "The Food-Energy-Nexus and Ecosystems", concentrating on energy-feeding strategies and their result on ecosystems and agricultural supply, was hosted by the UNU (Emodi et al., 2017).

However, the Bonn (2011) conference titled "The Water, Energy, and Food Security Nexus–Solutions for the Green Economy" remained a breakthrough such that the idea of the FEW nexus was noticed in the global programme (Hoff, 2011). Right from the conference, nearly 300 establishments across the globe inaugurated FEW nexus ambitions between 2011 and 2015 (Endo et al., 2017), and the number of FEW nexus journal articles has started growing historically. Figure 5.3 presents 530,185 research articles identified and retrieved between 2000 and 2023 based on the popular ScienceDirect database.

It is noticeable in Figure 5.3 that several articles were recognised and published with more concentration on food-energy-water rather than the FEW nexus until 2011, when the nexus of FEW started increasing. Concurrently, the three-branched FEW networks have steadily drawn researchers' attention, and research publications in this area have risen suddenly. Several studies on the FEW nexus focus on international, transborder, and national measures.

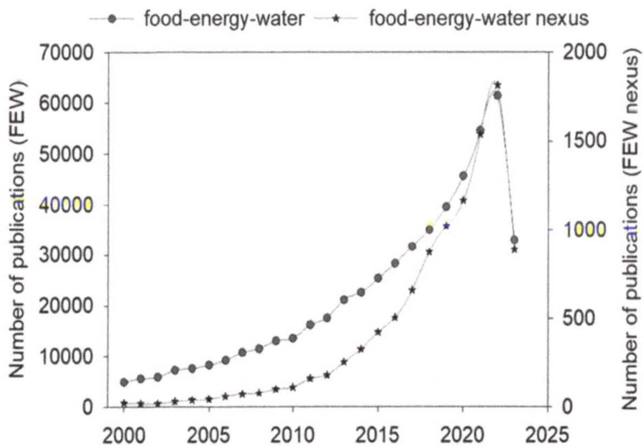

FIGURE 5.3 Description of the FEW and FEW nexus documents from 2000 to 2023.

5.2 STATE OF THE ART OF FOOD-ENERGY-WATER NEXUS

Presently, research on FEW nexus optimisations is relatively scarce. Most researchers have assumed optimisation as a separate resource instead of the optimisation of FEW as a complete resource. With regards to an individual resource or two resource optimisations, the research work performed by Javadinejad et al. utilised diverse means and algorithms for water resource optimisations (Wang et al., 2015; Javadinejad, Ostad-Ali-Askari and Eslamian, 2019). In contrast, Ju and Su et al. presented a multiple-design procedure for energy optimisation (Wang et al., 2015; Ju et al., 2016). Also, Guo and Cai separately developed algorithms for food structure optimisation and establishment, aiming to improve the differences in demand (Guo and Zhao, 2017; Shen et al., 2022). In another study, Cao et al. (2011) carried out optimisation at the limitation of water resources to design and establish a business in arid regions. The studies by Zhen et al. (2020) controlled hybrid interval two-stage fuzzy reliability coding algorithms to sustain interactions of the energy-water scheme organisation. Likewise, Chen et al. (2020) studied the nexus of the energy-water algorithm, an optimisation system with double risk aversion that can produce healthy optimisation results.

However, considering the complete optimisation of food, energy, and water, Zhang improved the water and energy application, and food generation was centred on the complete assessment process. In the same vein, Karan et al. (2018) proposed a model and designed a mathematical algorithm for predicting demand and generation, including food, energy, and water. The study by Fan et al. (2019) established a combined assessment of the FEW nexus for two parts with various stages of development through several linear regressions in a simultaneous equation method. Also, Li et al. (2019) proposed an optimisation algorithm aiming to exploit financial profits with negligible environmental effects. At the same time, Peng (2017) optimised the correlations among food, energy and water channels by initiating the theory of agreement.

5.2.1 Food-Energy-Water Nexus Viewpoints in Africa as a Case Study for a Developing Country

The FEW nexus is studied in various regions. For example, climate change is assessed in Africa from a fresh perspective based on studies undertaken through thc FEW nexus. It is based on many factors, including resource deficiencies, the high price of resources, and input costs, which might threaten the security at the grassroots level. The public's creativity, well-managed governance, and natural resources might secure FEW growth in the community, while encouraging sustainability and lowering domestic liabilities (Davidson et al., 2003). In sub-Saharan Africa, the research was based on climate change's consequences on food supply, energy generation, and water management (Misra, 2014). The result indicates that several rich natural resources in Africa, such as mineral and energy resources, are presently in the framework of the African condition. Nevertheless, their mismanagement is either totally cogent or entirely irrational, and the materials are frequently migrated out of Africa. In 1990, the region lacked infrastructural development and organisational and managerial competencies and had a decreasing societal growth rate, all of which led to a series of challenges.

Africa relies on agricultural products not only for local utilisation but also for export incomes as well. Unfortunately, agriculture is prone to deforestation, erosion, and the affordability of international markets. The area requires international support and growth strategies focused on FEW security (Bian and Liu, 2021). Even if the viewpoint is meagre, the region will nevertheless transform. In Northeast Africa, for example, Ethiopia is advancing in agricultural innovation. The energy supplies to the country are mainly from biomass, which, in the context of climate change, are assessed in Africa's hydroelectric power services built by the government. There is interference between water and agricultural services, and energy supply occurs in the country, as in Asia (Karlberg et al., 2015). There are different challenges in the locality of Lake Tana, Ethiopia, because of the increase in agricultural products and the rise in demand for energy uses such as transport, storage, and irrigation.

Hence, to solve this challenge, migrating to hydroelectricity becomes an alternative. This is because the waste produced by the agricultural activity can be used to generate bioenergy; this condition can generate synergy or competition between sectors of FEW. Although lots of water is used for bioenergy and agriculture, the country lacks enough to deliver (Mayor et al., 2015). Other African countries are equally encountering this issue. Climate change affects different African cities, such as Bulawayo in Zimbabwe, Cape Town in South Africa, Dar-es-Salam in Tanzania, and Cairo in Egypt. The water supplies for the production of hydroelectricity are in danger because of these challenges, and agricultural activities are present in those cities as well. Moreover, in different areas, the power given to the local government institutions needs to be increased, and the financial limitations coupled with the lack of technical knowledge to discourse climate change equally add up to the challenges in implementing the FEW nexus. The result brings a broad gap between the higher financial capabilities and the underprivileged, with the rise in food and energy prices and limited access to clean water (Chirisa and Bandauko, 2015).

On the other hand, South Africa follows a team of international communities for food price increments. Presently, 60% of South Africa's families are challenged with some levels of food uncertainty (Gulati et al., 2013). In 2007 and 2008, the cost of power rose by approximately 24%, upsetting all financial areas, including all customers. The Producer Price Index (PPI) shows that water and gas prices were up by around 96%, and electricity prices rose by approximately 177% in a year interval. Meanwhile, the prices of other food industries and processing materials have risen equally, while food prices have also been increased. Because of the increase in energy prices, water supply, and agricultural outputs, transportation and delivery are undesirably increased. Consequently, the situation increased the cost of water by 60%.

Regarding food production and safety, increasing water prices have caused greater difficulties because of water scarcity (Gondhalekar and Ramsauer, 2017). Several new strategies have evolved to solve South Africa's nexus problems. Some studies aim to improve moisture retention in grasslands and enhance nitrogen supply, reducing fertiliser use and increasing the efficiency of irrigation systems. Suitably, rainwater quality being affected by dung could be applied to produce water-soluble fertilisers and biogas, with the former utilised for domestic uses (Gondhalekar and Ramsauer, 2017). Generally, water use is a serious environmental and social resource issue in the African region. Other resource usage and climate change stress the ecosystem as the population of the Nile basin increases. The extra pressure leads to internal conflicts, with countries neighbouring the Nile fully involved in significant productive riparian regions. Africa has the Nile at the top among the Essal river schemes, and its Blue Nile tributary produces a total annual income of about 60% (Allam and Eltahir, 2019). At the Blue Nile River corner, Allam et al. generated the FEW nexus to exploit all the advantages by proficiently tackling water resources and land. The method researched, in this case, can assist in concealing parts of the basin with rain-fed agriculture, refining the soil in the system. Consequently, the yearly algorithm performed at the Blue Nile River water (Keskinen et al., 2015) produces savings of about 7.55 km^3.

Moreover, among the 11 irrigation systems that were designed from the master project in Ethiopia, three are projected to be beneficial. Water stored for hydropower generation arises at the cost of possible rain-fed agriculture in the basin. The collaboration of the countries in the area aims to promote effective agriculture and distribute the benefits and expenses. Hence, the transaction will result in collaboration in other regions (Keskinen et al., 2015). To progress further, the minor schemes require implementation at a greater level to assist the country and their counterparts. Water-use policies, simulated water commerce, technology, and transboundary supervisory of water resources are gaining ground to struggle with the FEW nexus challenges. As a result of this, the nexus system becomes essential to enable long-standing approaches and policies for food security and lasting sustainability in Africa.

5.2.2 Food-Energy-Water Nexus Viewpoints in Europe as a Case Study for a Developed Country

One of the major difficulties in Europe regarding the FEW nexus is equating the energy costs for agriculturalists, while authorising them to pay for irrigation costs. Also, supplying water to smaller societies is a core issue, particularly for hydropower, and making it known to the community is needed, too (Mayor et al., 2015). Uncertainty has become a repeated argument in the Global Economic Forum administration. Consequently, European countries must implement concepts and plans to avert disaster, improve assets, and prevent problems related to the security of the World Economic Forum. The idea of nexus regarding the World Economic Forum shows complicated interactions between investors and areas. However, few countries have strategies and policies prepared to solve the FEW nexus. For instance, the United Kingdom lacks strategies that impact the FEW nexus relationships, encouraging balances across FEW systems and infirming the capacity of the nation to resist climate change (Sharmina et al., 2016). Even with the latest improvements in the EU in recognising nexus challenges, a majority that produces the reliable basis of the policies is at sector-explicit, challenging average- and long-standing issues regarding resource administration and climate change adaptation. Science, dynamism, multidisciplinarity, and adaptability should be the focus of recent strategies and policymaking. The study undertaken in conjunction with UK professionals suggests that evaluation and examination in the framework of the FEW nexus demand a multi-investor and collaborative method across the information and policy-making procedures. This approach requires the dynamic involvement of investors from every segment (Bian and Liu, 2021).

The increase in global requests for natural resources in Europe, accompanied by climate change, has generated considerable problems in urban advancement and viability. For instance, among Germany's main cities, Munich is highly dependent on food importation. In this regard (Gondhalekar and Ramsauer, 2017), green agriculture can utilise vertical agriculture to grow viable food production. The vertical crops might generate 66% of Munich's vegetables and fruits, which is nearly the average requirements of the city. Moreover, many advantages of vertical agriculture can be observed, including the reduction of heat islands, which assists in lower energy utilisation by providing cooling during summer. Likewise, wastewater treatment and re-utilisation may cause water reserves to equal 26% when mixed with rainwater storage. Since the water stored might be used for vertical agriculture in urban farming, the plan for water consumption has been preserved. It can utilise the combined FEW nexus procedures to expand urban viability besides FEW security by resolving the challenges faced by FEW resource investors. A civilised and comprehensive transmission structure permits contributors to decrease symbiotic dangers. On the other hand, the aftermath of COVID-19 on several sectors of the FEW nexus and the accomplishment of the SDGs were also examined.

5.3 OPTIMISATION ANALYSIS

5.3.1 Food Optimisation for Sustainable Development

The frequent challenges in food have led to suggestions to prevent the additional increase of land utilised for agriculture and to grow products on lands where crop yields are low but efforts are being made to enhance resource contribution capacity, such as adding fertilisers and using irrigation systems (Aschemann-Witzel et al., 2019). Also, food waste and misuse should be reduced to increase the food utilisation

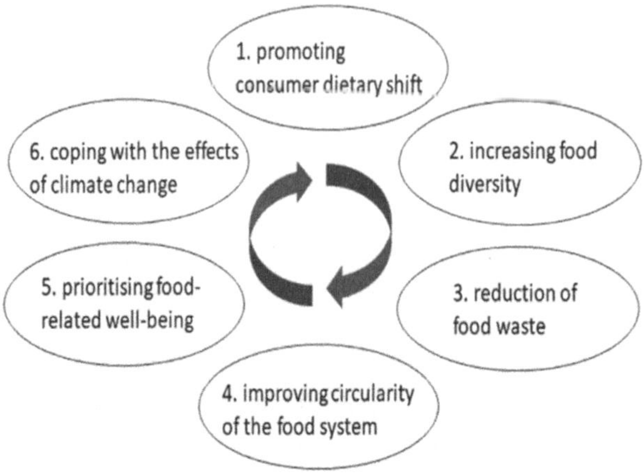

FIGURE 5.4 The six essential development revolutions towards a sustainable food sector (Aschemann-Witzel et al., 2019).

ratio and a change in interest from meat consumption to generating additional food for societal use while crops are utilised for livestock consumption (Foley et al., 2011). Meanwhile, Keating et al. (2014) organized the proposals into 14 so-called "food wedges," grouping them by activities to justifiably accommodate the demand for food predicted for 2050, organised into three major sets. However, these three sets are "reducing food demand," including meat demand and overfeeding, decreasing misuse, and moving biofuels to renewables without competing with food. The next is "filling the production gap," such as an increase in supply, products, and proficiency, and finally, "stopping wastes in generation possibility," for example, conservation against dilapidation, dangers to supply by virus and resistance, and influences of climate change moderation (Keating et al., 2014). Figure 5.4 shows a classification in the food sector for a required sustainable transformation.

Moreover, in 2016, the Global Network Against Food Crises collaborated with humanitarian and development organisations to prevent, formulate, and proffer solutions to global food crises. This association tries to decrease susceptibilities related to severe hunger, attain food security, enhance nutrition, and encourage sustainable agriculture and food systems, using a 3×3 approach as presented in Figure 5.5.

This method includes working at the global, regional, and national levels to sustain collaborations within the present systems and to optimise support, policy-making, strategy, and programming using a measurable approach.

5.3.2 Energy Optimisation for Sustainable Development

Energy systems are complicated structures where the generation and consumption should be harmonised properly. This is because there are several energy sources and few transformation technologies available for their development. For energy sustainability, diffusing renewable energy resources should be a top priority. Several research works were executed in power systems deliberating on wind, natural gas,

FIGURE 5.5 The 3 × 3 method to addressing food challenges (Food Security Information Network (FSIN) and Global Network against Food Crises (GNAFC), 2022).

and solar as a means of power supply. Komušanac et al. (2016) analysed the effect of more solar PV production and wind infiltrations on the nation's load power system research in Croatia. The result depicted a high share of wind power plants and installed solar PV capacities of 1.65 GW and 1.6 GW, respectively, as a means of sustainability.

Similarly, Zeng et al. (2016) executed a fixed-state combined electric power system and natural gas with reversing energy transformation. The projected system was authenticated/established by the IEEE-9 assessment with the 7-node natural gas system. The findings were convincing, and the correctness was satisfactory. Additionally, the time series of power load and wind power is utilised to examine the moderation result of the combined energy network. Eventually, the impact of power demand and wind power on power gas output and gas-fired energy production will also be examined. Prebeg et al. (2016) studied the long-term energy development of Croatian power systems through multi-purpose optimisation, concentrating on renewable energy and electric vehicle integrations.

In another study, Falke et al. (2016) studied a multi-purpose advancement and simulation approach to build a large-scale circulated energy network with over 100 houses. The established system is used in a region of an average-sized town in Germany to examine the impacts of various productivity analyses concerning full prices and releases of equivalent carbon (IV) oxide (CO_2). The Pareto-efficient techniques, technologies, and proficient solutions that can contribute more resourcefully to decreasing GHG releases are recognised.

5.3.3 Water Optimisation for Sustainable Development

The means to safe water, cleanness, and sanitation remain the fundamental human necessities for well-being and healthy living (Okampo and Nwulu, 2020). It is predicted that billions of people will suffer from water scarcity in 2030, unless improvement quadruples. The water demand is increasing because of urbanisation, quick

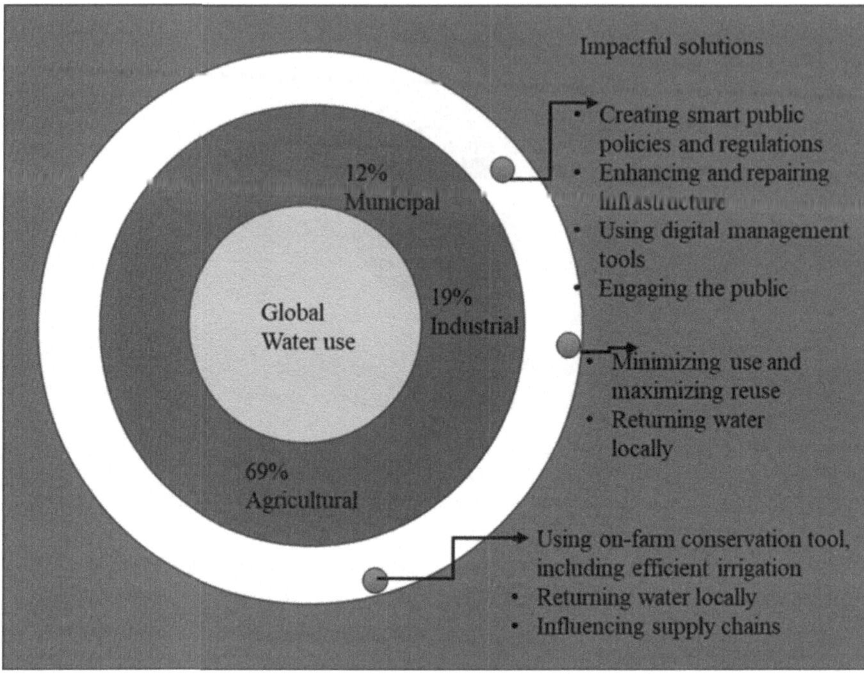

FIGURE 5.6 Global effective water use by the energy, industrial, and agricultural sectors.

growth in population, and growing water demands from energy, industry, and agriculture. Figure 5.6 demonstrates effective water utilisation by the three major sectors: energy, industrial, and agricultural.

To attain the global supply of consumable water, hygiene, and sanitation by 2030, the present pace of growth needs to increase four times. As a result, accomplishing these goals would save 829,000 inhabitants yearly, whose deaths might result from illnesses directly caused by unclean water, insufficient hygiene, and poor sanitation practices. However, to accelerate the efforts employed in addressing water security problems, the United Nations General Assembly pronounced 2018–2028 as the Water Action Decade for Sustainable Development (WADSD). This agrees with and matches the 2030 Programme for Sustainable Development Goals (SDGs 2015–2030). This agenda is noticed halfway through the Water Action Decade and the SDG period. This account—started by the United Nations University Institute for Water Environment and Health (UNU IWEH), the UN's think tank on water—offers an initial measurable international evaluation that estimates the condition of water security for 7.78 billion individuals in 186 nations.

5.3.4 Food-Energy-Water Optimisation for Sustainable Development

It has become imperative that one of the agendas in international research organisations is to optimise FEW by increasing the food supply as well as access to energy and water to revitalise rural people out of poverty (Bizikova et al., 2013). In line with this aim, they have funded energy and water for rural inhabitants, resulting in ineffective utilisation and dilapidation of resources. With regard to state programmes for socio-economic growth in rural communities (by funding energy and water), it is important to confirm that the resource base is continued, GHG releases are controlled, and water-borne diseases are checked.

Therefore, international bodies are deliberating on the FEW nexus—which discourses the relationship between the three essential resources to improve daily incomes and encourage socio-economic growth—when undertaking development activities (Newell et al., 2019). This is frequently delayed because various administrative agencies control these resources. However, integrated concerned agencies are trying to strategise and devise progress schemes, but these plans are usually not viable in the short to medium term. Therefore, it is necessary to determine methods to optimise the effectiveness of these three resources. The following possibilities are suggested for FEW optimisations:

1. Governments and development ministries should integrate food, energy, and water in their scheduling for socio-economic expansion.
2. Executive bodies for food, energy, and water should ensure a close relationship with higher authorities, including the economics or developmental departments, which might have a wider knowledge and order.
3. Implement the common food, energy, and water ethics into the plan design and application.
4. While developing fresh projects, it is necessary to analyse methods to combine the present government enterprises with the fresh project. For instance, assume a government parastatal is eager to develop high-voltage distribution systems to reduce outflows and supply dependable electricity to farmers to supply groundwater. Therefore, supplying constant electricity will relieve farmers' worries about irrigating crops. The farmers may extract more water than an aquifer can recharge, leading to unsustainable groundwater utilisation. Nevertheless, if a high-voltage distribution system is integrated with drip irrigation and optional irrigation facilities, water and electricity will be applied professionally and sustainably.

5.4 METHODOLOGY

An SLR performed on educational systems helps researchers to create sophisticated outline data or information on a particular study area (Chandler and Hopewell, 2013). Thus, this type of review offers a systematic method for classifying, separating, and combining a group of published research papers, journals, conferences, and books in search of sufficient proof related to particular research objectives or questions.

5.4.1 Bibliometric Analysis

The bibliometric study has often been utilised for systematic analysis of scientific publications and documentation to explore information in a certain area of study, and the analysis was first utilised in 1968 by Alan Pritchard (Chen et al., 2019). It is a statistical research technique. Furthermore, this research applied CiteSpace software as an tool for advancing information domain visualisation and bibliometric examination. CiteSpace assists in the temporal and structural studies of hybrid node kinds, networks, and cluster sights on scientific papers (Chen, 2006).

5.4.2 Impact Factor and Social Network Analysis

Impact factor (IF) is an essential parameter applied in the bibliometric study and is among the frequently used parameters to assess the relative effect of journals (Chiu and Ho, 2007). It is a consistent parameter designed through publications and citations of an article for the past two years (Garfield, 2006; Buela-Casal and Zych, 2012). Similarly, the h-index is another parameter of separate research accomplishment from quantity and quality domains, proposed by Hirsch in 2005 (Hirsch and Buela-Casal, 2014). On the other hand, social network examination is employed to recognise the networks among specific bodies and reproduce the importance of particulars and the strength or intensity of bonds with statistics (Newman, 2001). In the network graph, collaborations and individuals are characterised by bulked lines and nodes, respectively, and the nodes perform significant functions in the organisational connections. The role of various countries and organisations can be assessed by evaluating research partnerships with the advancement of network models and the computer's processing abilities.

5.4.3 Database, Search Terms, and Search Procedure

The article did a critical assessment and detailed investigation of appropriate research journals, books, and conferences from the databases of Web of Science, Scopus, and ScienceDirect, using the keywords "food," "energy," "water," and "nexus" in the title, abstract, and keywords of the database articles. During the search process, English was only used to examine peer-reviewed journal articles published from 2010 to May 2024. The search led to 417 documents, which were retrieved on 20 May 2023. However, the search methods' limitations were also considered by using alternative words, including electricity, power, agriculture, hydrolysis, and irrigation, to search for related articles in the title, abstract, and keywords to acquire 24 extra papers. Hence, an overall of 641 articles were recorded. Even though the papers chosen were restricted to peer-reviewed articles, tools, and procedures for nexus uses were initially established by intergovernmental or non-governmental studies and policy organisations.

5.4.3.1 Article Inclusion and Exclusion Criteria

Three criteria were chosen in selecting the articles, and they include:

1. The articles clearly use the idea of nexus regarding the sustainability of natural resources.
2. The articles expressively comprise the three resource areas: food, energy, and water.
3. All the articles propose or test detailed systematic tools for assessing the nexus.

Henceforth, if articles passed just the first two conditions but failed to propose or test systematic tools, the situation was considered as "conceptual." Out of the 641 articles presented, 50, representing 7.8%, were categorised as "conceptual" and 146 articles, representing 22.7%, were categorised as "methodological." The excluded articles were 147, representing 45.8%. To confirm that the review process was reliable, articles were excluded once they failed to pass single-selection conditions. Take, for instance, some articles that were unable to meet up with the sustainability of natural resources, but "nexus" was included as a tinkle term. Even though all three terms indicated were common, articles were excluded once resource nexus interactions between two terms were noticed. Perhaps, articles should concentrate mainly on water–energy relationships. In this case, the emphasis is placed on addressing the three-branched nexus because food, energy, and water systems are integrally more complicated compared to the two-branched nexus and the systematic methods for evaluating the FEW nexus are required in studying this extra intricacy. Lastly, since the key purpose of this research is to sustain extra advancement of healthy nexus approaches, articles that failed to propose or test exact systematic tools for assessing the nexus were equally excluded. The subsections of 73 "methodological" papers were coded for material connected to publication (e.g., country, keywords, year, discipline, and journal) and nexus approaches (e.g., quantitative, single discipline, and interdisciplinary). Similarly, the assessment included:

1. Whether the papers demanded a "new" technique, uniting several approaches in a discipline, or applying a single technique.
2. Whether the papers presented a favourite for a specific area.
3. Whether the articles had case studies.
4. If the articles had an aim or end purpose.
5. The scale of the study was all about.
6. Main limitations or challenges in using the nexus technique(s).

Meanwhile, a quantitative study was conducted regarding the drifts and features of nexus approaches described in the literature (e.g., applications of quantitative methods, number of publications over time, and variety of disciplines). All 50 "conceptual" articles were reviewed to collect the exact data gaps, study requirements, and prescriptive qualities connected to nexus methods and tools. These prescriptive qualities of nexus approaches presented in the literature are termed "key features." The key features represent the required qualities, for example, how techniques assist in accessing nexus study, what data and information are excluded or included, how interdisciplinary relationships are explained, and who is included in the study. In the same vein, the normative attributes of the nexus approach or study requirements from the "methodological" papers, where appropriate, were collected.

5.5 RESULTS AND DISCUSSION

SLR is an essential technique to clarify the uncertainty regarding the authenticity of a topic in connection with the key aim of describing some matters in an organised and detailed manner. The social network analysis used in this research supports

TABLE 5.1
Publications of Top 10 Subject Categories

Web of Science Categories	Record Count	% of 236
Environmental Sciences	153	64.831
Green Sustainable Science Technology	51	21.61
Water Resources	42	17.797
Environmental Studies	40	16.949
Engineering Environmental	33	13.983
Energy Fuels	15	6.356
Engineering Civil	14	5.932
Meteorology Atmospheric Sciences	14	5.932
Engineering Chemical	11	4.661
Geosciences Multidisciplinary	11	4.661

generating visualisations and analyses of various comparative techniques that assist research hypotheses. The social network analysis aims to produce clear, complicated relationships regarding the FEW nexus. However, Table 5.1 illustrates the top 10 published articles based on the subject category. Out of the 236 publications from Web of Science, articles (156), reviews (66), proceeding papers (11), and book chapters (8) were the main contributions that recorded about 66.4% of the overall publications. Other published documents recorded approximately 34%, such as abstracts and editorial materials. The top productive category is Environmental Sciences, accounting for 64.831% of all publications, followed by Green Sustainable Science Technology, accounting for 21.61%, and Water Resources, accounting for 17.797%. The journals in the subject categories of environmental sciences and water resources are the most influential.

5.5.1 Most Frequently Cited Papers and Features of Publications

The most frequently cited articles can reproduce the hot spots of research attention in FEW nexus studies. Since newly published articles may need more suitable citations for their rapid publication time, utilising the most cited article per annum for assessment purposes is sensible as presented in Table 5.2. A total of 50 concepts regarding the FEW nexus were recognised in the papers studied by ScienceDirect. They applied various techniques to examine a similar matter: food, water, and energy governance. For example, the highly cited articles with a total yearly citation of 264 and an average per year of 44 were published in *Review of Geophysics* in 2018, entitled "The global food-energy-water nexus." Its major contribution is the need to prioritise nutrition, not just food, in examining the nutritional significance of various climate and management schemes. The second top from the list has 167 and an average of 27.83 citations and is entitled "Understanding and managing the food-energy-water nexus - opportunities for water resources research." It was also

TABLE 5.2
Hierarchy of Articles According to Their Annual Citations

Title of the Article	References	Journal Title	Publication Year	Total Citations	Average per Year
The global food-energy-water nexus	D'Odorico et al. (2018)	*Reviews of Geophysics*	2018	264	44
Understanding and managing the FEW nexus – opportunities for water resources research	Cai et al. (2018)	*Advances in Water Resources*	2018	167	27.83
Improving the sustainability of organic waste management practices in the food-energy-water nexus: A comparative review of anaerobic digestion and composting	Lin et al. (2018)	*Renewable & Sustainable Energy Reviews*	2018	166	27.67
Agrivoltaics provide mutual benefits across the food-energy-water nexus in drylands	Wilberforce et al. (2019)	*Nature Sustainability*	2019	160	32
Food waste and the food-energy-water nexus: A review of food waste management alternatives	Kibler et al. (2018)	*Waste Management*	2018	158	26.33
Food-energy-water (FEW) nexus for urban sustainability: A comprehensive review	Zhang, Zhang, Chang, et al. (2019)	*Resources Conservation and Recycling*	2019	151	30.2
The food-energy-water nexus: Transforming science for society	Scanlon et al. (2017)	*Water Resources Research*	2017	148	21.14
Sustainability in the food-energy-water nexus: Evidence from BRICS (Brazil, the Russian Federation, India, China, and South Africa) countries	Ozturk (2015)	*Energy*	2015	117	13
An urban system framework to assess the trans-boundary food-energy-water nexus: implementation in Delhi, India	Ramaswami et al. (2017)	*Environmental Research Letters*	2017	94	13.43
A food-energy-water nexus approach for land use optimisation	Nie et al. (2019)	*Science of the Total Environment*	2019	90	18
Designing integrated local production systems: A study on the food-energy-water nexus	Hang et al. (2016)	*Journal of Cleaner Production*	2016	83	10.38

(Continued)

TABLE 5.2 (CONTINUED)
Hierarchy of Articles According to Their Annual Citations

Title of the Article	References	Journal Title	Publication Year	Total Citations	Average per Year
Environmental sustainability issues in the food-energy-water nexus: Breakfast cereals and snacks	Jeswani, Burkinshaw and Azapagic (2015)	*Sustainable Production and Consumption*	2015	78	8.67
Food-energy-water nexus: A life cycle analysis on virtual water and embodied energy in food consumption in the Tamar catchment, UK	Salmoral and Yan (2018)	*Resources Conservation and Recycling*	2018	73	12.17
A 40-year review of food-energy-water nexus literature and its application to the urban scale	Newell, Goldstein and Foster (2019)	*Environmental Research Letters*	2019	72	14.4
Wastewater treatment and reuse in urban agriculture: exploring the food, energy, water, and health nexus in Hyderabad, India	Miller-Robbie, Ramaswami and Amerasinghe (2017)	*Environmental Research Letters*	2017	70	10
The food-energy-water nexus and urban complexity	Romero-Lankao, McPhearson and Davidson (2017)	*Nature Climate Change*	2017	70	10
Wavelength-selective solar photovoltaic systems: Powering greenhouses for plant growth at the food-energy-water nexus	Loik et al. (2017)	*Earth's Future*	2017	69	9.86
Impacts of irrigated agriculture on food-energy-water-CO_2 nexus across meta coupled systems	Xu et al. (2020)	*Nature Communications*	2020	67	16.75
Food-energy-water nexus: Quantifying embodied energy and GHG emissions from irrigation through virtual water transfers in food trade	Vora et al. (2017)	*ACS Sustainable Chemistry & Engineering*	2017	66	9.43
Optimising resource use efficiencies in the food-energy-water nexus for sustainable agriculture: From conceptual model to decision support system	Tian et al. (2018)	*Current Opinion in Environmental Sustainability*	2018	62	10.33

SDG synergy between agriculture and forestry in the food, energy, water and income nexus: reinventing agroforestry?	van Noordwijk et al. (2018)	*Current Opinion in Environmental Sustainability*	2018	61	10.17
Environmental sustainability in the food-energy-water-health nexus: A new methodology and an application to food waste in a circular economy	Slorach et al. (2020)	*Waste Management*	2020	49	12.25
Emerging challenges and opportunities for the food-energy-water nexus in urban systems	Heard et al. (2017)	*Current Opinion in Chemical Engineering*	2017	49	7
Assessment of the features of coupling interaction of the food-energy-water nexus in China	Han, Yu and Cao (2020)	*Journal of Cleaner Production*	2020	48	12
Integrated scenarios to support analysis of the food-energy-water nexus	Van Vuuren et al. (2019)	*Nature Sustainability*	2019	47	9.4
Quantifying the urban food-energy-water nexus: The case of the Detroit metropolitan area	Liang et al. (2018)	*Environmental Science & Technology*	2019	47	9.4
Integrated governance for the food-energy-water nexus – The scope of action for institutional change	Märker, Venghaus and Hake (2018)	*Renewable & Sustainable Energy Reviews*	2018	46	7.67
Considering agricultural wastes and ecosystem services in food-energy-water-waste nexus system design	Garcia, Lovett and You (2019)	*Journal of Cleaner Production*	2019	45	9
Urban food-energy-water nexus indicators: A review	Arthur et al. (2019)	*Resources Conservation and Recycling*	2019	43	8.6
Uncovering urban food-energy-water nexus based on physical input-output analysis: The case of the Detroit metropolitan area	Feng et al. (2019)	*Applied Energy*	2019	43	8.6
Stakeholder analysis for the food-energy-water nexus in Phoenix, Arizona: Implications for nexus governance	White et al. (2017)	*Sustainability*	2017	42	6
System dynamics simulation of regional water supply and demand using a food-energy-water nexus approach: Application to Qazvin Plain, Iran	Naderi et al. (2021)	*Journal of Environmental Management*	2021	41	13.67
Biorefineries and the food, energy, water nexus – Towards a whole systems approach to design and planning	Martinez-Hernandez and Samsatli (2017)	*Current Opinion in Chemical Engineering*	2017	41	5.86

(Continued)

TABLE 5.2 (CONTINUED)
Hierarchy of Articles According to Their Annual Citations

Title of the Article	References	Journal Title	Publication Year	Total Citations	Average per Year
An extended environmental input-output lifecycle assessment model to study the urban food-energy-water nexus	Sherwood, Clabeaux and Carbajales-Dale (2017)	*Environmental Research Letters*	2017	41	5.86
Measuring urban food-energy-water nexus sustainability: Finding solutions for cities	Yuan, Chiueh and Lo (2021)	*Science of the Total Environment*	2021	40	13.33
A metropolitan scale water management analysis of the food-energy-water nexus	Guan et al. (2020)	*Science of the Total Environment*	2020	39	9.75
Towards productive cities: Environmental assessment of the food-energy-water nexus of the urban roof mosaic	Toboso-Chavero et al. (2019)	*Journal of Industrial Ecology*	2019	38	7.6
Food-energy-water-waste nexus systems optimisation for New York State under the COVID-19 pandemic to alleviate health and environmental concerns	Zhao and You (2021)	*Applied Energy*	2021	34	11.33
Identifying critical supply chains: An input-output analysis for food-energy-water nexus in China	Xiao et al. (2019)	*Ecological Modelling*	2019	34	6.8
Including nature in the food-energy-water nexus can improve sustainability across multiple ecosystem services	Hanes, Gopalakrishnan and Bakshi (2018)	*Resources Conservation and Recycling*	2018	32	5.33
A bibliometric analysis of food-energy-water nexus literature	Opejin et al. (2020)	*Sustainability*	2020	31	7.75
Understanding the tele-coupling mechanism of urban food-energy-water nexus: Critical sources, nodes, and supply chains	Zhang, Zhang, Hao, et al. (2019)	*Journal of Cleaner Production*	2019	30	6
Advancing the food-energy-water nexus: Closing nutrient loops in arid river corridors	Mortensen et al. (2016)	*Environmental Science & Technology*	2016	29	3.63
Gateway to the perspectives of the food-energy-water nexus	Proctor, Tabatabaie and Murthy (2021)	*Science of the Total Environment*	2021	27	9

Integrating embedded resources and network analysis to understand food-energy-water nexus in the US	Mahjabin et al. (2020)	*Science of the Total Environment*	2020	27	6.75
A conceptualisation of the urban food-energy-water nexus sustainability paradigm: Modeling from theory to practice	Schulterbrandt Gragg et al. (2018)	*Frontiers in Environmental Science*	2018	27	4.5
Quantifying direct and indirect spatial food-energy-water (FEW) nexus in China	Liang et al. (2020)	*Environmental Science & Technology*	2020	26	6.5
Developing a design-led approach for the food-energy-water nexus in cities	Roggema and Yan (2019)	*Urban Planning*	2019	25	5
Methane emissions from artificial waterbodies dominate the carbon footprint of irrigation: A study of transitions in the food-energy-water-climate nexus (Spain, 1900–2014)	Aguilera et al. (2019)	*Environmental Science & Technology*	2019	24	4.8
Optimising food-energy-water (FEW) nexus to foster collective resilience in urban-rural systems	Sukhwani et al. (2019)	*Progress in Disaster Science*	2019	24	4.8

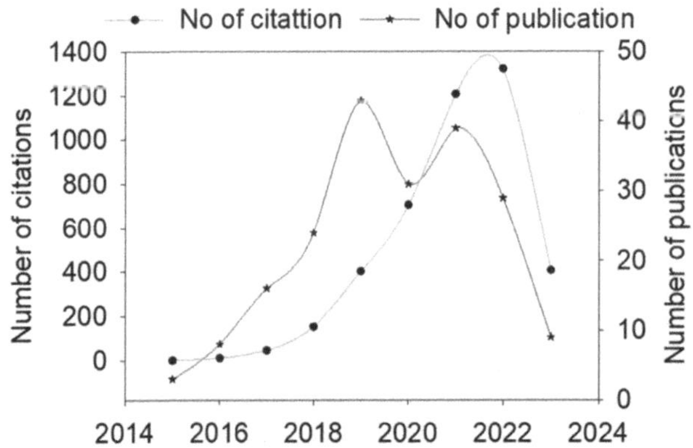

FIGURE 5.7 The number of citations and publications from 2014 to May 2023.

published in 2018 and elaborated on the opportunities in FEW nexus management. The third and fourth articles were "Renewable & Sustainable Energy Reviews" and "Nature Sustainability," published in 2018 and 2019, respectively. While the third article has 166 and an average of 27.67, the fourth article has 160 and an average of 32 citations.

Among the publications presented in Table 5.2, there are only two articles in each of 2015 and 2016, but the yearly publication rose steadily with a sharp increase in 2019. The average publication length changes fairly, with a total average of 12.78 pages. The average number of citations per article grew from 3.63 in 2016 to 44 in 2018, disregarding the years with fewer articles (2015 and 2016). The number of authors is equally increasing from 1 in 2015 to 236 in 2021, with the average number of authors per article of 4.72. Figure 5.7 shows the number of citations and publications from 2015 to 2024.

Hence, these features show that the FEW nexus is a hot spot research field with great interest from academia and experts.

5.5.2 Countries Mapping

Some countries have devoted more attention to the new research approach than others. The development of overlay visualisations assisted researchers in visualising countries committed to FEW nexus optimisations. The unit of study was on countries, while bibliography coupling was based on the analysis method. Five were set out as a country's threshold requirement for this chapter, and 57 out of 103 countries met the thresholds. Most of the citations were found in the United States, recording 12,608. The next country was the United Kingdom, with values of 6,101 citations, while Germany and China recorded 4,521 and 4,154 citations, respectively. Similarly, the United States provided the published articles accounting for 355 papers, seconded by China, which recorded 224 articles. The United Kingdom and Germany recorded 157 and 95 articles, respectively. The country's impact in advancing the recent research field is based on the number of articles published, overall link strength, and the number of citations recorded. The overall strength of the correlation

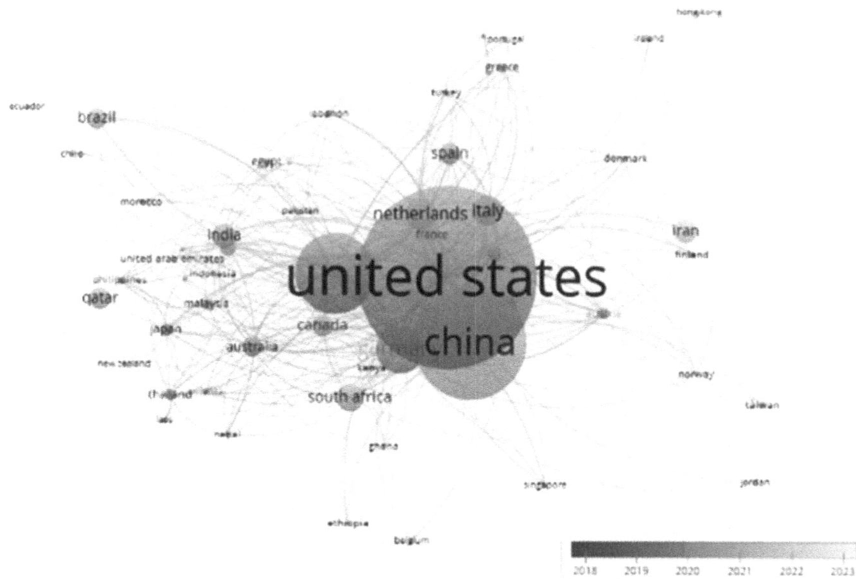

FIGURE 5.8 Density visualisation among countries in food-energy-water influence.

portrays the extent to which the published articles of a certain nation impacted the published articles of the other nations covered in this research. As a result, the under-listed countries were regarded as possessing the most effect on FEW nexus optimisations. Figure 5.8 presents the countries' relationships and an overlay visualisation of countries linked by citations. The size of the circle shows the country's commitment to the area. Moreover, the overlay graph depicts that the nations with the highest contribution had a more visual outlook.

The research trends in Figure 5.8 describe that from 2018 to 2020, FEW nexus optimisations were studied mainly by countries such as New Zealand, the United Kingdom, Egypt, Australia, India, Mexico, South Korea, France, Japan, and the Netherlands. However, the top three countries in recent research interest from 2021 to 2024 are the United States, China, and Germany, with total link strengths recorded as 327, 187, and 146, respectively. This graphical illustration of the contributing nations will support prospective researchers in creating scientific relationships, building mutual project collaborations, and participating in fresh methods.

5.5.3 Keyword Mapping

The keywords are indispensable in research areas because they support recognising and deliberations on the subject being researched. In this chapter, keyword analysis was studied, and the "kind of analysis" was found as "co-occurrence," while the "unit of analysis" was ascertained as "all keywords." A threshold number was limited to 50 times to ensure that no keywords were utilised less than 50 times. Because of these limitations, it was noticed that only 135 from a total of 2,581 keywords passed the threshold for inclusion. The result depicts the frequently occurring keywords: water,

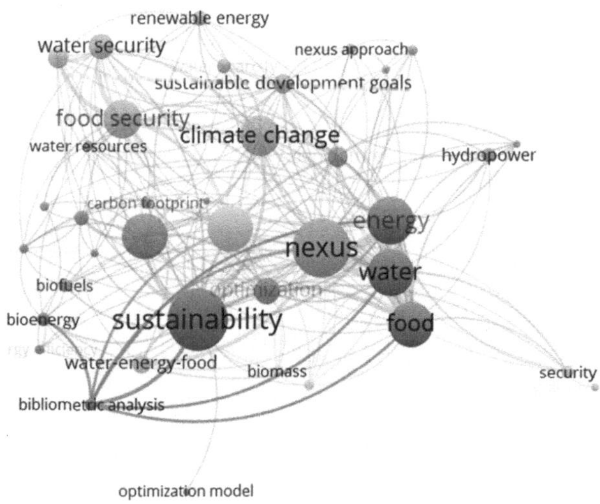

FIGURE 5.9 Network visualisation describing the relationship among keywords in the FEW nexus.

food, energy, nexus, and climate change. The network of visualisation of the co-occurrence of keywords is presented in Figure 5.9, with their correlation. Even though the keyword node location shows how frequently a word occurs in articles, the size also indicates the frequency of occurrence in those articles. Moreover, the result depicts that some keywords have larger nodes than others, showing their relevance in the FEW nexus research area.

According to network analysis in Figure 5.9, the term "optimisation" is not a focal point from the primary keywords gathered. The result shows that the idea of FEW nexus optimisations remains under-researched. Also, from the result obtained, the predominant study areas are "water-energy nexus", or "energy-water nexus", "food-energy-water nexus", and "water-food." Due to the FEW-centred trend in most research, some scholars have condemned the network's recent examinations as inadequate and subdivided. The study on energy-water, water-food, or food-energy nexuses alone weakens the authentic plan of advancing specific interdisciplinary viewpoints to succeed in traditional areas. The decision has been reached that different keywords could be clearly differentiated in the network to show their co-occurrence in various research works. Hence, Table 5.3 shows the co-occurrence of the analysis of clusters of keywords.

The prevailing keywords emerge in the red cluster and occur as the highest number of keywords, comprising 15 keywords. As a result, this cluster produces an area with the highest research concentration in FEW nexus optimisations.

5.5.4 A Nexus Optimisation Approach

As development continues to spin worldwide, an effort is needed to uphold the nexus generation based on community study with the government, private sectors, municipalities, and scholars. Nexus-based studies may generate new perceptions into

TABLE 5.3
Clusters of Keywords Co-occurrence Mapping

Colour of Clusters	Selected Keywords	No of Keywords	Comments
Red	Bibliometric analysis, bioenergy, bioenergy, biofuels, carbon footprint, decision-making, energy efficiency, food waste, food-energy-water nexus, network analysis, optimisation, optimisation model, sustainability, water footprint, water resources, water stress	15	The interest of the researchers is highest in the field of FEW nexus optimisation
Green	Climate change, energy security, food security, renewable energy, sustainable development, water scarcity, water security	7	The interest of the researchers is highly focused on the field of FEW nexus optimisation
Blue	Agriculture, ecosystem, energy, food, hydropower, water	6	Researchers focus moderately on the field of FEW nexus optimisation
Yellow	Biomass, integration, sustainable development	3	The interest of researchers is less on the field of FEW nexus optimisation
Pink	Food-energy-water, nexus approach, transdisciplinary	3	The interest of researchers is less on the field of FEW nexus optimisation
Light blue	Policy integration, security	2	The least interest is observed from the researchers in the field of FEW nexus optimisation
Orange	Nexus, water-energy-food	2	The least interest is observed from researchers in the field of FEW nexus optimisation.

possible deals and alliances of investments in infrastructure, guidelines, or reserve management opportunities.

5.5.4.1 Accessibility to Water Quality and Energy Application in Rural Agricultural Sectors

One of the essential and prospective research areas comprises cost evaluation, techno and socio-economic advantages of fresh drinking water analysis, and its circulatory set-up and water waste administration. Providing solutions caused by poor water quality on societal health remains a serious global concern, and nexus agendas can provide the main perception. The current water quality challenges are better accountable to the ground and surface water systems by water testing, which could aid in future examination by administrative or scientific involvement. The benefits of the perspectives obtained through the FEW nexus system and future advancement could enhance means to reliable and safe drinking water, particularly by encouraging residents on hygiene and community health programmes in rural settings. These aims are obtainable through organising agricultural policing with an energy and water advancement agenda and encouraging the accountable utilisation of pesticides and fertilisers to avoid extra water pollution. This promotes financial inducements to reliable water organisation, changes water-concentrated technologies with competent ones, and controls the release of unprocessed industrial wastes. Moreover, systems modelling that connects hydrologic simulation devices that attract the spatiotemporal supply of water through financial means that reproduce the prices and advantages of water distributions can be utilised to assess socio-economic effects through a series of freshwater treatment remedies. Such contexts can reproduce energy-linked charges and necessities of water treatment and supply organisations that help in installing modern energy projects such as the present renewable energy development strategies.

5.5.4.2 Irrigation and Energy Accessibility

Access to irrigation and energy could be another main nexus approach in reaching smallholder farmers in rural areas to enhance their livelihood. While the total volume of water on the earth's surface is about 70%, the yearly spatial supply of the resources can restrict the chances of increasing the means of crop irrigation. Additionally, smallholder farmers may have no finance to advance in farm- or public-based irrigation distribution systems, including ground or surface water drainage or storage facilities. To this end, international collaborations could aid investors in comprehending the possible financial worth of increasing access to irrigation while pondering the prospective prices of abridged water availability and the energy needs of the aforesaid schemes. A combined examination may assist in recognising critical spatial areas for energy sustenance and access to irrigation. Such reserves are possibly a significant adjustment in climate change mechanisms for agriculturists, though more study is required to assist in categorising structural reserves and rules.

5.5.4.3 The Effects of Renewable Energy

Presently, the world is striving for energy accessibility through renewable energy development strategies, and more study is required to comprehend how the said plans can relate to food or water distribution schemes. Increasing access to energy may enable irrigation growth, reducing worries regarding the security of local food, while intensifying agriculture might increase the concerns for quality water. Moreover, rules benefiting biomass or biofuel electricity as a remedy for renewable energy might add extra pressure on the execution of agriculture in the global communities, which may worsen local water rivalry. Systems modelling and involvement with resident investors could assist in comprehending and analysing these main nexus networks and balances.

5.5.4.4 Enterprises and Encouragements in Local Regions

The local communities are always neglected whenever opportunities and creativity are demanded. Presently, few or no authorised FEW nexus enterprises exist in local communities. The National Action Plan for Climate Change (NAPCC) was the closest enterprise reported in Spanish and established in conjunction with non-government and governmental organisations to organise diligence on climate change adjustment and moderation for effective utilisation of food, energy, and water topics. Although there was no clarity of policies indicating the FEW nexus, it identified the networks of connectivity in each field. Similarly, the FEW nexus enterprises in the local region that were backed up by the Deutsche Gesellschaft für Internationale Zusammenarbeit and the EU formed the Caribbean and Latin American unit in alliance with the Economic Commission for Latin America and the Caribbean (ECLAC) to grow healthy conversations relating to FEW nexus challenges in local policies. Furthermore, fiscal and financial encouragement might be another requirement to kickstart investigations on enterprises and related schemes to nexus matters. Generally, study approaches that concurrently discourse the nexus might assist in forming plans that solidify the resilience of the FEW nexus in the global community.

5.5.5 Aftermath of COVID-19 on the FEW Nexus

The outbreak of COVID-19 has had a huge impact on the FEW nexus and interrupted the balance between the food, energy, and water sectors (Barbier and Burgess, 2020). The effect can be categorised into direct and indirect impacts (Yin et al., 2022). Direct impacts began from the consequences of circulation or infection of the virus into food and water supplies, whereas indirect impacts originated from the environmental and socio-economic consequences of lockdown measures (Yin et al., 2022). Following the inception of the outbreak, there was a notable surge in water usage because of sensitive hygiene necessities, like sanitation efforts and handwashing associated with COVID-19 treatment materials and fumigation of containment areas (Bellie, 2021). The increase in water consumption extended to both domestic and medical sectors, reflecting a change in water demand patterns compared to pre-outbreak levels (Al-Saidi and Hussein, 2021). This upsurge in consumption of water resources resulted from the lockdown brought about by stay-at-home and recreational activities, such as gardening (Karan et al., 2018). On the other hand, industrial water consumption was reduced meaningfully because of postponed processes within lockdown measures (Al-Saidi and Hussein, 2021).

The energy sector experienced the same challenges, with unstable demands perceived as an instant effect of COVID-19 (Jiang, Van Fan and Klemeš, 2021). Previously, the industrial sectors mostly controlled the increased demand for electricity. Hence, industrial closures resulted in a large cut in electricity demand, though global energy demands finally recovered as lockdown limits were reduced to revive industrial and economic activities (Tsao et al., 2021). Similarly, the pandemic affected the renewable energy sector mostly because of the interruption of incentives and subsidies (Tsao et al., 2021). Agricultural sectors faced interruptions in food accessibility and distribution, particularly for marginalised populations in rural areas (Dobson et al., 2020).

According to studies, COVID-19 left local inhabitants under critical socio-economic pressure due to their restrictions in accessing native food sources (Ruiz-Salmón et al., 2021; Rai et al., 2022). Notwithstanding, global food consumption upsurged during the outbreak, intensifying differences in food access and distribution (Eftimov et al., 2020). Movement limitations caused undesirable effects on farmer livelihoods, employment, and local incomes, also deepened by agricultural export interruptions (Yu et al., 2020). These interruptions triggered across the food supply chain resulted in wastage and shortages. Besides, the COVID-19 outbreak interrupted advancement towards attaining the SDGs, mostly those connected to food security (SDG 2), water resources (SDG 6), and energy resilience (SDG 7) (Ioannou and Laspidou, 2023). The interconnection of the FEW nexus with SDGs emphasises the extensive effects of the outbreak on global sustainability efforts.

5.6 CONCLUSION

The present systematic review shows that up till now, the study on FEW nexus optimisations remains at an early stage, without an established method for solving the challenges encountered. This chapter was systematically reviewed, focusing on data retrieved from the ScienceDirect, Scopus, and Web of Science databases. During the screening and review process, 258 published papers were retrieved from Web of Science, 355 papers from Scopus, and 50 articles were procured from ScienceDirect. Publications from the Web of Science revealed that only 66 papers were review articles, and none clarified the optimisation of the FEW nexus. In addition to the low reports of this review paper, several authors addressed themes and topics within the FEW nexus frameworks and neglected optimisation. Although there were few review articles, the number of citations and publications for the FEW nexus has steadily improved. Many articles concentrate on environmental sciences, which record 64.831%, green sustainable science technology accounting for 21.61%, and water resources recording 17.797%. The number of citations increased nationally, with the United States as the most productive country recording 12,608 citations. The United Kingdom and Germany are next, accounting for 6,101 and 4,521 citations, respectively. Meanwhile, the United States, China, and the United Kingdom recorded 355, 224, and 157, respectively, as the most published papers. The result shows that research collaborations among countries and organisations are gaining widespread interest. The studies often cited articles and keywords that discovered the study hotspots and trends. Therefore, with this growing research focus on the FEW nexus, it is necessary to establish a better and complete knowledge of its challenges globally. For instance, how to concurrently discourse challenges in water quality and distribution, enhance food security through extended means of irrigation, and supply stable electricity in local sectors. These will need "systems-focused" methods that clearly identify important networks between segments. The integrated tactics regularly disclose interactions or imperfections in or even upset the conservative understanding of relationships between food supply, water resources, and energy generation. Consequently, they provide a better knowledge of policymaking and generate ingenious results. The recent COVID-19 outbreak greatly affected food security, energy resilience, and water resources. Hence, the harmful effects of COVID-19 on the food, energy, and water sectors forced restrictions on the realisation of the SDGs.

REFERENCES

Agbor, M. E. et al. (2023) 'Potential impacts of climate change on global solar radiation and PV output using the CMIP6 model in West Africa'. *Cleaner Engineering and Technology*, 13, p. 100630.

Aguilera, E. et al. (2019) 'Methane emissions from artificial waterbodies dominate the carbon. footprint of irrigation: A study of transitions in the food–energy–water–climate nexus (Spain, 1900–2014)'. *Environmental Science & Technology*, 53(9), pp. 5091–5101.

Ajanovic, A. (2011) 'Biofuels versus food production: Do biofuels production increase food prices?' *Energy*, 36(4), pp. 2070–2076.

Albrecht, T. R., Crootof, A. and Scott, C. A. (2018) 'The water-energy-food nexus: A systematic review of methods for nexus assessment'. *Environmental Research Letters*, 13(4), p. 43002.

Allam, M. M. and Eltahir, E. A. B. (2019) 'Water-energy-food nexus sustainability in the Upper Blue Nile (UBN) Basin'. *Frontiers in Environmental Science*, 7, p. 5.

Al-Rousan, N., Isa, N. A. M. and Desa, M. K. M. (2018) 'Advances in solar photovoltaic tracking systems: A review'. *Renewable and Sustainable Energy Reviews*, 82, pp. 2548–2569. Doi: 10.1016/j.rser.2017.09.077

Al-Saidi, M. and Hussein, H. (2021) 'The water-energy-food nexus and COVID-19: Towards a systematization of impacts and responses'. *Science of the Total Environment*, 779, p. 146529.

Apeh, O. O., Meyer, E. L. and Overen, O. K. (2022) 'Contributions of solar photovoltaic systems to environmental and socioeconomic aspects of national development—A review'. *Energies*, 15(16), p. 5963.

Apeh, O. O., & Nwulu, N. I. (2025). Machine learning approach for short-and long-term global solar irradiance prediction. *Journal of Environmental & Earth Sciences*, 7(01).

Aquastat, F. A. O. (2011) *FAO's information system on water and agriculture*. FAO, Food and Agriculture Organization of the United Nations.

Arthur, M. et al. (2019) 'Urban food-energy-water nexus indicators: A review'. *Resources, Conservation and Recycling*, 151, p. 104481.

Aschemann-Witzel, J. et al. (2019) 'A sense of sustainability? – How sensory consumer science can contribute to the sustainable development of the food sector'. *Trends in Food Science & Technology*, 90, pp. 180–186.

Barbier, E. B. and Burgess, J. C. (2020) 'Sustainability and development after COVID-19'. *World Development*, 135, p. 105082.

Bellie, S. (2021) 'COVID-19 and water'. *Stochastic Environmental Research and Risk Assessment*, 35(3), pp. 531–534.

Bian, Z. and Liu, D. (2021) 'A comprehensive review on types, methods and different regions related to water–energy–food nexus'. *International Journal of Environmental Research and Public Health*, 18(16), p. 8276.

Birol, E. and Das, S. (2010) 'Estimating the value of improved wastewater treatment: The case of River Ganga, India'. *Journal of Environmental Management*, 91(11), pp. 2163–2171.

Bizikova, L. et al. (2013) *The water-energy-food security nexus: Towards a practical planning and decision-support framework for landscape investment and risk management*. International Institute for Sustainable Development Winnipeg.

Buela-Casal, G. and Zych, I. (2012) 'What do the scientists think about the impact factor?' *Scientometrics*, 92(2), pp. 281–292.

Cai, X. et al. (2018) 'Understanding and managing the food-energy-water nexus–opportunities for water resources research'. *Advances in Water Resources*, 111, pp. 259–273.

Cao, X. et al. (2011) 'Planting structure optimization in the arid area with constrained water resources: A case study of Korla Xinjiang'. *Resources Science*, 33, pp. 1714–1719.

Chandler, J. and Hopewell, S. (2013) 'Cochrane methods-twenty years experience in developing systematic review methods'. *Systematic Reviews*, 2(1), pp. 1–6.

Chen, C. (2006) 'CiteSpace II: Detecting and visualizing emerging trends and transient patterns in scientific literature'. *Journal of the American Society for Information Science and Technology*, 57(3), pp. 359–377.

Chen, C. et al. (2020) 'Planning energy-water nexus systems based on a dual risk aversion optimization method under multiple uncertainties'. *Journal of Cleaner Production*, 255, p. 120100.

Chen, D. et al. (2019) 'Recent progress on the water–energy–food nexus using bibliometric analysis'. *Current Science*, 117(4), pp. 577–586.

Chhipi-Shrestha, G., Hewage, K. and Sadiq, R. (2017) 'Water–energy–carbon nexus modeling for urban water systems: System dynamics approach'. *Journal of Water Resources Planning and Management*, 143(6), p. 4017016.

Chirisa, I. and Bandauko, E. (2015) 'African cities and the water-food-climate-energy nexus: An agenda for sustainability and resilience at a local level'. In *Urban forum* (pp. 391–404). Springer.

Chiu, W.-T. and Ho, Y.-S. (2007) 'Bibliometric analysis of tsunami research'. *Scientometrics*, 73(1), pp. 3–17.

D'Odorico, P. et al. (2018) 'The global food-energy-water nexus'. *Reviews of Geophysics*, 56(3), pp. 456–531.

David, L. O. et al. (2022) 'Integrating fourth industrial revolution (4IR) technologies into the water, energy & food nexus for sustainable security: A bibliometric analysis. *Journal of Cleaner Production*, 363, p. 132522.

Davidson, O. et al. (2003) 'The development and climate nexus: The case of sub-Saharan Africa'. *Climate Policy*, 3(sup1), pp. S97–S113.

Desa, U. N. (2018) *The sustainable development goals report 2018*. United Nations.

Dobson, A. P. et al. (2020) 'Ecology and economics for pandemic prevention'. *Science*, 369(6502), pp. 379–381.

Eftimov, T. et al. (2020) 'COVID-19 pandemic changes the food consumption patterns'. *Trends in Food Science & Technology*, 104, pp. 268–272.

Emodi, N. V. et al. (2017) 'Energy policy for low carbon development in Nigeria: A LEAP model application'. *Renewable and Sustainable Energy Reviews*, 68, pp. 247–261.

Endo, A. et al. (2017) 'A review of the current state of research on the water, energy, and food nexus'. *Journal of Hydrology: Regional Studies*, 11, pp. 20–30.

Falke, T. et al. (2016) 'Multi-objective optimization and simulation model for the design of distributed energy systems, *Applied Energy*, 184, pp. 1508–1516.

Fan, C., Lin, C.-Y. and Hu, M.-C. (2019) 'Empirical framework for a relative sustainability evaluation of urbanization on the water–energy–food nexus using simultaneous equation analysis'. *International Journal of Environmental Research and Public Health*, 16(6), p. 901.

Feng, C. et al. (2019) 'Uncovering urban food-energy-water nexus based on physical input-output analysis: The case of the Detroit Metropolitan Area'. *Applied Energy*, 252, p. 113422.

Foley, J. A. et al. (2011) 'Solutions for a cultivated planet'. *Nature*, 478(7369), pp. 337–342. Doi: 10.1038/nature10452

Food Security Information Network (FSIN) & Global Network against Food Crises (GNAFC). (2022, May). *Global Report on Food Crises 2022*. https://www.fsinplatform.org/report/global-report-food-crises-2022

Garcia, D. J., Lovett, B. M. and You, F. (2019) 'Considering agricultural wastes and ecosystem. Services in food-energy-water-waste nexus system design'. *Journal of Cleaner Production*, 228, pp. 941–955.

Garcia, D. J. and You, F. (2016) 'The water-energy-food nexus and process systems engineering: A new focus'. *Computers & Chemical Engineering*, 91, pp. 49–67.

Garfield, E. (2006) 'The history and meaning of the journal impact factor'. *JAMA*, 295(1), pp. 90–93.

Gondhalekar, D. and Ramsauer, T. (2017) 'Nexus city: Operationalizing the urban water-energy-food nexus for climate change adaptation in Munich, Germany'. *Urban Climate*, 19, pp. 28–40.

Guan, X. et al. (2020) 'A metropolitan scale water management analysis of the food-energy-water nexus'. *Science of the Total Environment*, 701, p. 134478.

Gulati, M. et al. (2013) 'The water–energy–food security nexus: Challenges and opportunities for food security in South Africa'. *Aquatic Procedia*, 1, pp. 150–164.

Guo, Z. L. and Zhao, X. (2017) 'Evolution and structure optimization of grain and cash crops in Shandong province based on 1986–2015 statistics data'. *Chinese Journal of Agricultural Resources and Regional Planning*, 38, pp. 164–171.

Han, D., Yu, D. and Cao, Q. (2020) 'Assessment on the features of coupling interaction of the food-energy-water nexus in China'. *Journal of Cleaner Production*, 249, p. 119379.

Hanes, R. J., Gopalakrishnan, V. and Bakshi, B. R. (2018) 'Including nature in the food-energy-water nexus can improve sustainability across multiple ecosystem services'. *Resources, Conservation and Recycling*, 137, pp. 214–228.

Hang, M. Y. L. P. et al. (2016) 'Designing integrated local production systems: A study on the food-energy-water nexus'. *Journal of Cleaner Production*, 135, pp. 1065–1084.

Heard, B. R. et al. (2017) 'Emerging challenges and opportunities for the food–energy–water nexus in urban systems'. *Current Opinion in Chemical Engineering*, 17, pp. 48–53.

Heilig, G. K. (2012) 'World urbanization prospects: The 2011 revision'. *United Nations, Department of Economic and Social Affairs (DESA), Population Division, Population Estimates and Projections Section, New York*, 14, p. 555.

Helmstedt, K. J. et al. (2018) 'Innovating at the food, water, and energy interface'. *Journal of Environmental Management*, 209, pp. 17–22.

Hirsch, J. E. and Buela-Casal, G. (2014) 'The meaning of the h-index'. *International Journal of Clinical and Health Psychology*, 14(2), pp. 161–164.

Hoff, H. (2011) 'The water, energy, and food security nexus; Solutions for the green economy'. In *Proceedings of the Bonn 2011 Conference*, Bonn, Germany (pp. 1–52). Available at: https://www.sei.org/publications/understanding-the-nexus

Ioannou, A. E. and Laspidou, C. S. (2023) 'Cross-mapping important interactions between water-energy-food nexus indices and the SDGs'. *Sustainability*, 15(10), p. 8045.

IPCC and Team CW (2014) 'Climate change 2014: Synthesis report'. In Rajendra K. Pachauri, & Leo A. Meyer (Eds.), *Contribution of working groups I, II and III to the fifth assessment report of the intergovernmental panel on climate change* (p. 408). IPCC Geneva, Switzerland.

Javadinejad, S., Ostad-Ali-Askari, K. and Eslamian, S. (2019) 'Application of multi-index decision analysis to management scenarios considering climate change prediction in the Zayandeh Rud River Basin'. *Water Conservation Science and Engineering*, 4, pp. 53–70.

Jeswani, H. K., Burkinshaw, R. and Azapagic, A. (2015) 'Environmental sustainability issues in the food–energy–water nexus: Breakfast cereals and snacks'. *Sustainable Production and Consumption*, 2, pp. 17–28.

Jiang, P., Van Fan, Y. and Klemeš, J. J. (2021) 'Impacts of COVID-19 on energy demand and consumption: Challenges, lessons and emerging opportunities'. *Applied Energy*, 285, p. 116441.

Ju, L. et al. (2016) 'Multi-objective operation optimization and evaluation model for CCHP and renewable energy-based hybrid energy systems driven by distributed energy resources in China'. *Energy*, 111, pp. 322–340.

Karan, E., Asadi, S., Mohtar, R., & Baawain, M. (2018). Towards the optimization of sustainable food-energy-water systems: A stochastic approach. *Journal of Cleaner Production*, *171*, 662–674. https://doi.org/10.1016/j.jclepro.2017.10.051

Karlberg, L. et al. (2015) 'Tackling complexity: Understanding the food-energy-environment nexus in Ethiopia's Lake Tana sub-basin'. *Water Alternatives*, 8(1), pp. 710–734.

Keating, B. A. et al. (2014) 'Food wedges: Framing the global food demand and supply challenge towards 2050'. *Global Food Security*, 3(3–4), pp. 125–132.

Keskinen, M. et al. (2015) 'Water-energy-food nexus in a transboundary river basin: The case of Tonle Sap Lake, Mekong River Basin'. *Water*, 7(10), pp. 5416–5436.

Kibler, K. M. et al. (2018) 'Food waste and the food-energy-water nexus: A review of food waste management alternatives'. *Waste Management*, 74, pp. 52–62.

Komušanac, I., Ćosić, B. and Duić, N. (2016) 'Impact of high penetration of wind and solar PV generation on the country power system load: The case study of Croatia'. *Applied Energy*, 184, pp. 1470–1482.

Kulp, S. A. and Strauss, B. H. (2019) 'New elevation data triple estimates of global vulnerability to sea-level rise and coastal flooding'. *Nature Communications*, 10(1), pp. 1–12.

Lele, U., Klousia-Marquis, M. and Goswami, S. (2013) 'Good governance for food, water and energy security'. *Aquatic Procedia*, 1, pp. 44–63.

Levinson, A., Rosenberg, C. and Yansane, A. (1977) 'The political economy of energy and agriculture in the third world'. In William Lockeretz (Ed.), *Agriculture and energy* (pp. 639–655). Elsevier.

Li, M. et al. (2019) 'Stochastic multi-objective modeling for optimization of water-food-energy nexus of irrigated agriculture'. *Advances in Water Resources*, 127, pp. 209–224.

Liang, S. et al. (2018) 'Quantifying the urban food-energy-water nexus: The case of the Detroit metropolitan area'. *Environmental Science & Technology*, 53(2), pp. 779–788.

Liang, Y. et al. (2020) 'Quantifying direct and indirect spatial food–energy–water (FEW) nexus in China'. *Environmental Science & Technology*, 54(16), pp. 9791–9803.

Lin, L. et al. (2018) 'Improving the sustainability of organic waste management practices in the food-energy-water nexus: A comparative review of anaerobic digestion and composting'. *Renewable and Sustainable Energy Reviews*, 89, pp. 151–167.

Loik, M. E. et al. (2017) 'Wavelength-selective solar photovoltaic systems: Powering greenhouses for plant growth at the food-energy-water nexus'. *Earth's Future*, 5(10), pp. 1044–1053.

Mahjabin, T. et al. (2020) 'Integrating embedded resources and network analysis to understand food-energy-water nexus in the US'. *Science of the Total Environment*, 709, p. 136153.

Märker, C., Venghaus, S. and Hake, J.-F. (2018) 'Integrated governance for the food–energy–water nexus–The scope of action for institutional change'. *Renewable and Sustainable Energy Reviews*, 97, pp. 290–300.

Martinez-Hernandez, E. and Samsatli, S. (2017) 'Biorefineries and the food, energy, water nexus—Towards a whole systems approach to design and planning'. *Current Opinion in Chemical Engineering*, 18, pp. 16–22.

Mayor, B. et al. (2015) 'Application of a water–energy–food nexus framework for the Duero river basin in Spain'. *Water International*, 40(5–6), pp. 791–808.

Meadows, D. H., Meadows, D. L., Randers, J., & Behrens III, W. W. (1972). *The Limits to Growth: A Report for the Club of Rome's Project on the Predicament of Mankind.* Universe Books.

Miller-Robbie, L., Ramaswami, A. and Amerasinghe, P. (2017) 'Wastewater treatment and reuse in urban agriculture: Exploring the food, energy, water, and health nexus in Hyderabad, India'. *Environmental Research Letters*, 12(7), p. 75005.

Misra, A. K. (2014) 'Climate change and challenges of water and food security'. *International Journal of Sustainable Built Environment*, 3(1), pp. 153–165.

Mortensen, J. G. et al. (2016) 'Advancing the food-energy-water nexus: Closing nutrient loops. in arid river corridors'. *Environmental Science & Technology*, 50(16), pp. 8485–8496.

Moss, R. H. et al. (2008) *Towards new scenarios for analysis of emissions, climate change, impacts, and response strategies.* Pacific Northwest National Lab. (PNNL).

Naderi, M. M. et al. (2021) 'System dynamics simulation of regional water supply and demand using a food-energy-water nexus approach: Application to Qazvin Plain, Iran'. *Journal of Environmental Management*, 280, p. 111843.

Newell, J. P., Goldstein, B. and Foster, A. (2019) 'A 40-year review of food–energy–water nexus literature and its application to the urban scale'. *Environmental Research Letters*, 14(7), p. 73003.

Newell, P. et al. (2019) 'Climate-smart agriculture? Governing the sustainable development goals in sub-Saharan Africa'. *Frontiers in Sustainable Food Systems*, 3, p. 55.

Newman, M. E. J. (2001) 'Scientific collaboration networks. I. Network construction and fundamental results'. *Physical Review E*, 64(1), p. 16131.

Nie, Y. et al. (2019) 'A food-energy-water nexus approach for land use optimization'. *Science of the Total Environment*, 659, pp. 7–19.

Okampo, E. J. and Nwulu, N. I. (2020) 'Optimal energy mix for a reverse osmosis desalination unit considering demand response'. *Journal of Engineering, Design and Technology*, 18(5), pp. 1287–1303.

Opejin, A. K. et al. (2020) 'A bibliometric analysis of food-energy-water nexus literature'. *Sustainability*, 12(3), p. 1112.

Ozturk, I. (2015) 'Sustainability in the food-energy-water nexus: Evidence from BRICS (Brazil, the Russian Federation, India, China, and South Africa) countries'. *Energy*, 93, pp. 999–1010.

Peng, S. P. (2017) 'Synergistic optimization of water-energy-food in the Yellow River basin'. *Water Science and Technology*, 5, pp. 681–689.

Prebeg, P. et al. (2016) 'Long-term energy planning of Croatian power system using multi-objective optimization with a focus on renewable energy and the integration of electric vehicles'. *Applied Energy*, 184, pp. 1493–1507.

Proctor, K., Tabatabaie, S. M. H. and Murthy, G. S. (2021) 'Gateway to the perspectives of the food-energy-water nexus'. *Science of the Total Environment*, 764, p. 142852.

Rai, P. K. et al. (2022) 'The effects of COVID-19 transmission on environmental sustainability and human health: Paving the way to ensure its sustainable management'. *Science of the Total Environment*, 838, p. 156039.

Ramaswami, A. et al. (2017) 'An urban systems framework to assess the transboundary food-energy-water nexus: Implementation in Delhi, India'. *Environmental Research Letters*, 12(2), p. 25008.

Roggema, R. and Yan, W. (2019) 'Developing a design-led approach for the food-energy-water nexus in cities'. *Urban Planning*, 4(1), pp. 123–138.

Romero-Lankao, P., McPhearson, T. and Davidson, D. J. (2017) 'The food-energy-water nexus and urban complexity'. *Nature Climate Change*, 7(4), pp. 233–235.

Ruiz-Salmón, I. et al. (2021) 'The fishing and seafood sector in the time of COVID-19: Considerations for local and global opportunities and responses'. *Current Opinion in Environmental Science & Health*, 23, p. 100286.

Sachs, I. and Silk, D. (1990) *Food and energy: Strategies for sustainable development*. United Nations University Press.

Salmoral, G. and Yan, X. (2018) 'Food-energy-water nexus: A life cycle analysis on virtual water and embodied energy in food consumption in the Tamar catchment, UK'. *Resources, Conservation and Recycling*, 133, pp. 320–330.

Scanlon, B. R. et al. (2017) 'The food-energy-water nexus: Transforming science for society'. *Water Resources Research*, 53(5), pp. 3550–3556.

Schulterbrandt Gragg, R. et al. (2018) 'A conceptualization of the urban food-energy-water nexus sustainability paradigm: Modeling from theory to practice'. *Frontiers in Environmental Science*, 6, p. 133.

Sharmina, M. et al. (2016) 'A nexus perspective on competing land demands: Wider lessons from a UK policy case study'. *Environmental Science & Policy*, 59, pp. 74–84.

Shen, J. et al. (2022) 'Impacts, challenges and suggestions of the electricity market for hydro-dominated power systems in China'. *Renewable Energy*, 187, 743–759.

Sherwood, J., Clabeaux, R. and Carbajales-Dale, M. (2017) 'An extended environmental input–output lifecycle assessment model to study the urban food–energy water nexus'. *Environmental Research Letters*, 12(10), p. 105003.

Slorach, P. C. et al. (2020) 'Environmental sustainability in the food-energy-water-health nexus: A new methodology and an application to food waste in a circular economy'. *Waste Management*, 113, pp. 359–368.

Sukhwani, V. et al. (2019) 'Optimizing Food-Energy-Water (FEW) nexus to foster collective. resilience in urban-rural systems'. *Progress in Disaster Science*, 1, p. 100005.

Tian, H. et al. (2018) 'Optimizing resource use efficiencies in the food–energy–water nexus for sustainable agriculture: From conceptual model to decision support system'. *Current Opinion in Environmental Sustainability*, 33, pp. 104–113.

Toboso-Chavero, S. et al. (2019) 'Towards productive cities: Environmental assessment of the food-energy-water nexus of the urban roof mosaic'. *Journal of Industrial Ecology*, 23(4), pp. 767–780.

Troell, M. et al. (2014) 'Does aquaculture add resilience to the global food system?' *Proceedings of the National Academy of Sciences*, 111(37), pp. 13257–13263.

Tsao, Y.-C. et al. (2021) 'A risk-sharing-based resilient renewable energy supply network model under the COVID-19 pandemic'. *Sustainable Production and Consumption*, 25, pp. 484–498.

Uhlenbrook, S. and Connor, R. (2019) *The United Nations world water development report 2019: Leaving no one behind.* UNESCO.

Vahabzadeh, M., Afshar, A. and Molajou, A. (2023) 'Energy simulation modeling for water-energy-food nexus system: A systematic review'. *Environmental Science and Pollution Research*, 30(3), pp. 5487–5501.

van Noordwijk, M. et al. (2018) 'SDG synergy between agriculture and forestry in the food, energy, water and income nexus: Reinventing agroforestry?' *Current Opinion in Environmental Sustainability*, 34, pp. 33–42.

Van Vuuren, D. P. et al. (2019) 'Integrated scenarios to support analysis of the food–energy–water nexus'. *Nature Sustainability*, 2(12), pp. 1132–1141.

Vora, N. et al. (2017) 'Food–energy-water nexus: Quantifying embodied energy and GHG emissions from irrigation through virtual water transfers in food trade'. *ACS Sustainable Chemistry & Engineering*, 5(3), pp. 2119–2128.

Wang, B. et al. (2015) 'Urban water resources allocation under the uncertainties of water supply and demand: A case study of Urumqi, China'. *Environmental Earth Sciences*, 74, pp. 3543–3557.

White, D. D. et al. (2017) 'Stakeholder analysis for the food-energy-water nexus in Phoenix, Arizona: Implications for nexus governance'. *Sustainability*, 9(12), p. 2204.

Wilberforce, T. et al. (2019) 'Prospects and challenges of concentrated solar photovoltaics and enhanced geothermal energy technologies'. *Science of the Total Environment*, 659, pp. 851–861.

Wittfogel, K. A. (1955) 'Developmental aspects of hydraulic societies'. In Julian H. Steward (Ed.), *Irrigation civilizations: A comparative study* (pp. 43–52). Pan American Union.

Xiao, Z. et al. (2019) 'Identifying critical supply chains: An input-output analysis for food-energy-water nexus in China'. *Ecological Modelling*, 392, pp. 31–37.

Xu, Z. et al. (2020) 'Impacts of irrigated agriculture on food–energy–water–CO2 nexus across metacoupled systems'. *Nature Communications*, 11(1), p. 5837.

Yin, C. et al. (2022) 'Recover the food-energy-water nexus from COVID-19 under sustainable development goals acceleration actions'. *Science of the Total Environment*, 817, p. 153013.

Yu, X. et al. (2020) 'The impact of COVID-19 on food prices in China: Evidence of four major food products from Beijing, Shandong and Hubei Provinces'. *China Agricultural Economic Review*, 12(3), pp. 445–458.

Yuan, M.-H., Chiueh, P.-T. and Lo, S.-L. (2021) 'Measuring urban food-energy-water nexus sustainability: Finding solutions for cities'. *Science of the Total Environment*, 752, p. 141954.

Zeng, Q. et al. (2016) 'Steady-state analysis of the integrated natural gas and electric power system with bi-directional energy conversion'. *Applied Energy*, 184, pp. 1483–1492.

Zhang, P., Zhang, L., Chang, Y., et al. (2019) 'Food-energy-water (FEW) nexus for urban: A comprehensive review'. *Resources, Conservation and Recycling*, 142, pp. 215–224.

Zhang, P., Zhang, L., Hao, Y., et al. (2019) 'Understanding the tele-coupling mechanism of urban food-energy-water nexus: Critical sources, nodes, and supply chains'. *Journal of Cleaner Production*, 235, pp. 297–307.

Zhao, N. and You, F. (2021) 'Food-energy-water-waste nexus systems optimization for New York State under the COVID-19 pandemic to alleviate health and environmental concerns'. *Applied Energy*, 282, p. 116181.

Zhen, J. L. et al. (2020) 'Energy-water nexus planning of regional electric power system within an inexact optimization model in Tangshan City, China'. *Journal of Cleaner Production*, 266, p. 121997.

6 Managing the Food-Energy-Water-Ecosystem Nexus Method to Support Sustainable Development Goals through Renewable Energy

6.1 INTRODUCTION

Renewable energy is a significant prospect in improving resource management within the FEWE nexus. This energy system can address the nexus, which could precisely describe the relations between Agenda 2030 and climate change deliberations, as emphasised (Liu et al., 2018). Renewable energy technologies could confront various trade-offs between FEWE, significantly assisting every part (Apeh and Nwulu, 2024b). The benefits of renewable energy include saving energy and emissions, reducing dependence on fossil fuels, and providing better security. The decentralised nature of several renewable energy technologies presents an opportunity to develop energy and water availability, mostly in rural areas, thereby positively influencing food security. The IRENA emphasises the possibility of renewable energy in the FEWE nexus with a keen report (International Renewable Energy Agency (IRENA), 2022). The contribution of renewable energy in the global primary energy source is expected to rise from 16% in 2020 to 77% in 2050 in the 1.5°C Scenario, demanding a yearly progress contribution of 13 times the present share, as shown in Figure 6.1. This progress is projected to alleviate the primary energy supply because of improved energy efficiency and the rise in the implementation of renewables. Throughout this change, the energy mix will experience a deep shift, bringing about a net growth of 61% points in the contribution of renewable energy, motivated by a mix of end-use electrification, direct consumption, and renewable fuels. Reaching such a substantial level of penetration in renewable energy is essential to meeting the global climate points, necessitating considerable investments, policy reinforcement, and constant innovative energies.

Besides, the World Energy Outlook through the International Energy Agency investigates the function of renewable energy within the FEW nexus, explaining its

DOI: 10.1201/9781003633846-6

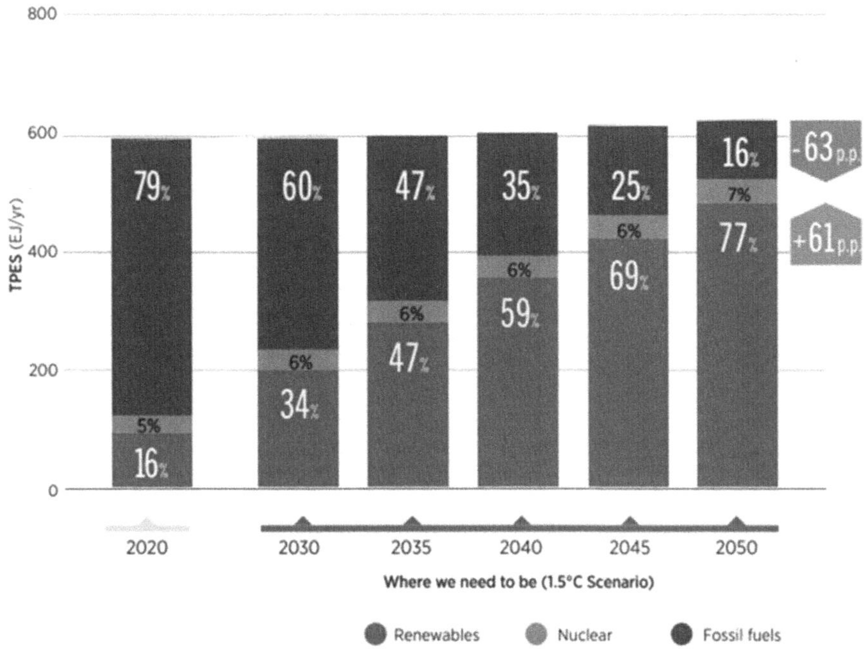

FIGURE 6.1 The total primary energy mix distribution from 2020 to 2050 under the 1.5°C scenario.

transformative potential (Gielen et al., 2019). With regard to the chosen technology, renewable energy could mitigate water supplies in energy generation. Notably, energy generation using wind and solar PV requires a minimal quantity of water, contributing to better efficiency. Geothermal and concentrated solar panels (CSPs) use heat while generating electricity, requiring more water. However, compared to conventional energy sources, renewable energy systems have been demonstrated to consume less water across all countries, as examined in the report presented by IRENA (Hassan et al., 2024). Several renewable energy resources, including wind, tidal, and solar, are easily accessible and need less water input, which benefits general water efficiency.

Renewable energy technologies are essential contributors to improving water security by enhancing availability, security, and affordability (Apeh and Nwulu, 2024a). This energy type could achieve energy needs with the water supply chain through solar-based pumping solutions, offering a lucrative substitute for diesel generators or on-grid electricity supply. Large water utilities gradually implement distributed renewable energy sources to offset electricity prices and decrease carbon footprints. In wastewater treatment, renewable energy provides solutions to energy demands and forms a nearby energy loop, employing the energy potential of wastewater through local bases, including biogas from waste. The combination of renewable energy presents economic opportunities in the food domain, reducing energy shortage losses and increasing supply chain productivity (Apeh and Nwulu, 2025).

Moreover, renewable energy could offer instant energy or be integrated through wide-range installations into the current traditional energy supply chain.

Geothermal heat and solar energy could be utilised in conjunction with waste energy as sources of dehydration energy in food processing. While recognising challenges regarding water utilisation and land use in bioenergy production, it has an encouraging role in the WEFE nexus and can be achieved under a definite limit (de Abreu and Machado, 2023). For example, installing anaerobic digesters on farms uses crop residues, along with food and animal waste, to generate a locally available source of electricity or heat. Integrating safety measures and guidelines is necessary for sustainable bioenergy operations. Furthermore, other renewable energy technologies could complement hydropower energy production, possibly lessening pressure on water resources and resolving trade-offs between water use and power generation. The nexus method is vital in recognising critical stress points associated with hydropower development. Furthermore, the nexus method can potentially reinforce the activities designed to attain the SDGs and offer specific contributions in several dimensions.

Firstly, the nexus method can increase growth by establishing complementary goals and aims that can be collectively accomplished. Interactions among goals and sectors offer insights into how individual aims can concurrently support various objectives. Secondly, the nexus method fosters partnerships and collaborations, playing a vital role in attaining the targets of the 2030 Agenda. Thirdly, it serves as a basis for producing solutions by investigating strategies from various entities or countries. On the broader framework, the nexus method becomes influential in progressing renewable energy distribution beyond the scope of SDG 7, which targets ensuring universal access to sustainable, reliable, affordable, and modern energy. For instance, applying the nexus method can disclose opportunities where renewable energy involvement contributes to accomplishing SDG 2, addressing hunger, guaranteeing food security, enhancing nutrition, and encouraging sustainable agriculture, as presented in Figure 6.2 (Segovia-Hernández, Contreras-Zarazúa and Ramírez-Márquez, 2023).

Furthermore, it could serve an essential role in growing SDG 6, ensuring the universal availability and sustainable management of water and sanitation. Additionally, the nexus method enables the incorporation of other SDGs, including 12, 13, and 15, expanding its effect across different SDGs.

6.2 DEVELOPING A CONCEPTUAL OUTLINE CONNECTING CLIMATE CHANGE WITH FOOD, ENERGY, WATER, AND ASSOCIATED ECOSYSTEM ACTIVITIES

Climate change, food, energy, water, and associated ecosystem activities are integrally related, wherein activities affecting one area always influence the others. Understanding the intricate and dynamic interaction among these components is important to efficiently tackle the rising demands for food, energy, freshwater, and sustenance within the challenges of climate change, escalating population growth, and compromised ecosystem resilience.

The conceptual outline that causes climate change in the environment includes temperature variations, changes in snowpack, and precipitation shift patterns. They have

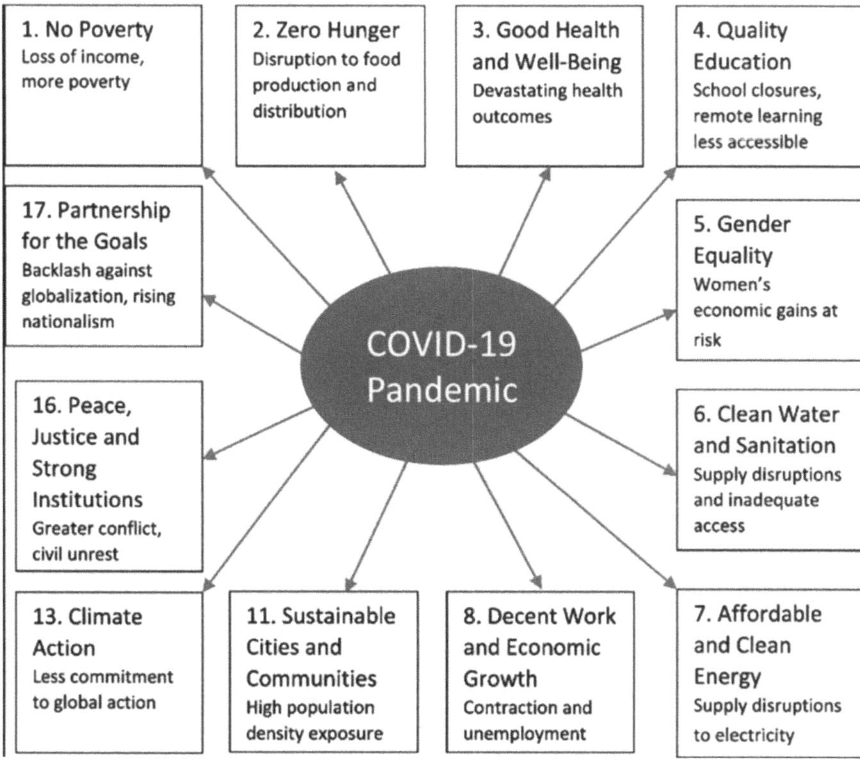

FIGURE 6.2 The impact of COVID-19 on the sustainable development goals (Barbier and Burgess, 2020).

definite directions associated with food, energy, water, and ecosystem procedures. The central ideas of this outline include:

1. The effect of climate change on precipitation, temperature, and snowpack, and its consequences on water resources like floods, stream flow dynamics, drought, surface water availability, water quality, watershed conditions, groundwater levels, and ecological processes.
2. The connectivity of these climate-change-induced effects with the life cycles of food, energy, and related ecosystem processes.

Figure 6.3 shows the complex and interrelated structural model, depicting how main climate change affects the life cycles of food, energy, water, and associated ecosystem processes. The effects of climate change initiators are described at the centre of this conceptual model, presenting their inclusive effects on the FEWE life cycles, causing higher conflicts and trade-offs within natural resource sectors.

Climate change initiators, including snowpack, precipitation, and temperature, meaningfully affect water production quality, quantity, and timing, subsequently

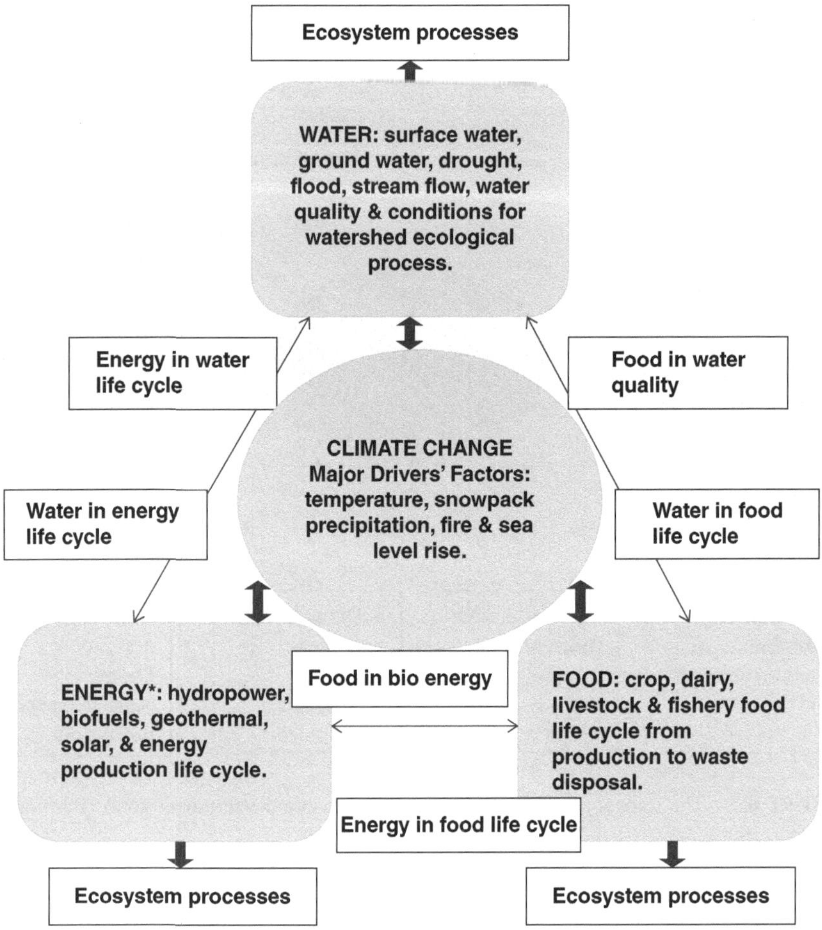

FIGURE 6.3 The complex structural model depicting how climate change affects the life cycles of food, energy, water, and associated ecosystem processes (Liu, 2016).

determining its consumption across food and energy production, agriculture, urban development, and efforts towards environmental conservation. The complex relationships and intricate connections exist in the following ways:

1. Climate change serves as a key initiator of snowpack reduction, rise in temperature, and precipitation fluctuations, resulting in regular floods and droughts that can reduce the availability of water and worsen water stress, thus affecting the life cycles of FEWE assistances obtained from water for wildlife and fish.
2. Water is a critical part of the energy life cycle for hydropower production processes. Similarly, energy is important for distribution, conveyance, water extraction, distribution, treatment, and heating for different agricultural and urban purposes.

3. Water consumption in food production includes livestock rearing, harvesting, irrigation, and processing, which greatly influences water resources and watershed ecosystems through agricultural runoff comprising pesticides, herbicides, fertilisers, and food waste.
4. Energy is needed for food production, distribution, and transportation, with crops and biomass serving as sources for biofuel production.
5. The production and consumption of FEW involve considerable effects on water and land resources, frequently resulting in conflicts with ecosystem processes, affecting species and habitats such as stream flow dynamics, wetlands, forests, water quality, and watersheds, mainly regarding wildlife and fish species. The knowledge of the intricate interrelations among these components is crucial for developing combined resource management plans that produce assistance, enable the assessment of trade-offs, and settle various interests and goals, thus allowing mitigation and adaptation efforts in response to climate change.

6.3 INTEGRATING FEW NEXUS WITH CLIMATE CHANGE

The FEW nexus is instrumental in solving various environmental challenges, the foremost of which is climate change (Liu, 2016; Al-Saidi and Elagib, 2017). Climate change is complicated by industrialisation, natural resource depletion, urbanisation, and economic expansion (Mohtar, 2022). Identifying the interrelated nature of these environmental challenges with climate change highlights the significance of using the FEW nexus to attain environmental sustainability (Bleischwitz, Johnson and Dozler, 2014). The failures of sector-driven management methods in tackling climate change also emphasise the necessity for the FEW nexus (Muller, 2015). Consequently, Al-Saidi and Elagib (2017) even projected that climate change could evolve as a critical driver for applying the FEW nexus (Muller, 2015).

To efficiently moderate and acclimatise to climate change, it is estimated that by 2030, energy resources will increase by 30%, production of food needs to grow by 50%, and an immediate 30% extra conservation of freshwater resources is required (Rai, 2023). Climate-change-induced extreme weather events, such as floods and droughts, present serious challenges for sustaining evolving economies. The increase in temperatures and changes in rainfall patterns because of climate change will greatly influence areas with inadequate financial inducements for resilience and adaptation (Niang et al., 2017). An important 2011 report from the World Economic Forum stressed the necessity for global leaders to study the interconnection of vital worldwide challenges (Waughray, 2011). The consequences of growing global temperatures cover food, water, and ecosystems. Mega-deltas, necessary for food production, experience sharp risks from increasing sea levels and flooding (Beddington, n.d.).

Tackling climate change needs a significant drop in global GHG emissions, aiming at a 50–60% reduction by 2050 concerning the present state, as demonstrated in Figure 6.4. The FEW nexus is recognised as a crucial actor in realising this determined goal (Beddington, n.d.). It is regarded as a socio-ecological systems method explicitly intended to encounter the challenges presented by climate change (Mabhaudhi et al., 2016). Present-day circumstances have a higher incidence of

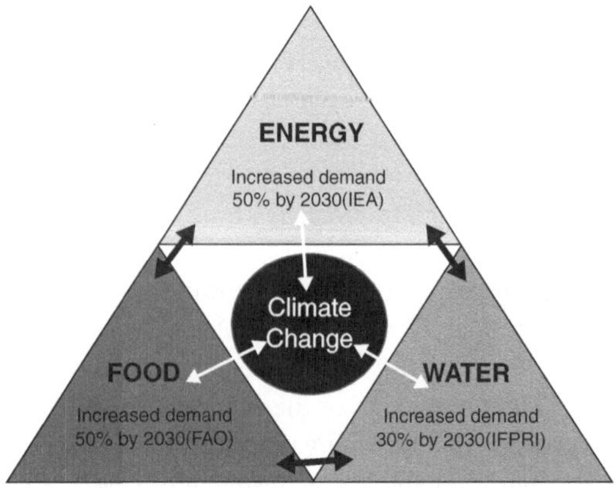

FIGURE 6.4 Demonstration of a perfect storm scenario (Rai, 2023).

extreme weather disasters, like floods and droughts, which can meaningfully impact agrarian-based economies (Rai et al., 2022).

The integration of renewable energy sources like biofuels is strongly connected with the FEW nexus and the challenges posed by climate change, as their production depends on natural resources to curtail GHG emissions (Chhandama and Rai, 2023). The study by Nzuma et al. (2010) presented numerous adaptation approaches for areas extremely vulnerable to climate change, such as (i) encouraging climate-smart agriculture, (ii) developing Early Warning Systems (EWS), (iii) executing combined water resource management, (iv) promoting the application of low-carbon renewable energies, and (v) increasing monitoring and modelling capacities across all areas of the FEW nexus (Nzuma et al., 2010). Besides, from the perception of the FEW nexus, enhanced management of transboundary natural resources can assist in reinforcing climate effects and accelerating climate action efforts. Thus, integrating the FEW nexus into environmental management strategies can strengthen energy flexibility and environmental sustainability, enabling effective climate action (Nzuma et al., 2010).

6.4 RENEWABLE ENERGY TRANSITION PATHWAYS

The change towards sustainable energy is a wide shift inspired by the overbearing need for environmental sustainability and urgent challenges concerning climate change, economic reformation, and local air pollution. Fundamentally, this change lies in the carbon footprint connected with our current energy usage. Currently, a considerable share of CO_2 emissions originates from our energy consumption, which has historically been powered by the combustion of fossil fuels to power transportation systems, homes, and industries. Nevertheless, this dependency comes at a substantial environmental price. Urban centres, where energy usage is huge, frequently face local air pollution, causing a decrease in air quality and unfavourably affecting public health, possibly decreasing life expectancy (Atlantic Council, 2021; Polcyn et al., 2023). Beyond local effects, the broader repercussions on the global climate are indisputable. In the atmosphere, GHG absorption has attained unparalleled levels, thereby instigating a global agreement on the need for a radical decrease.

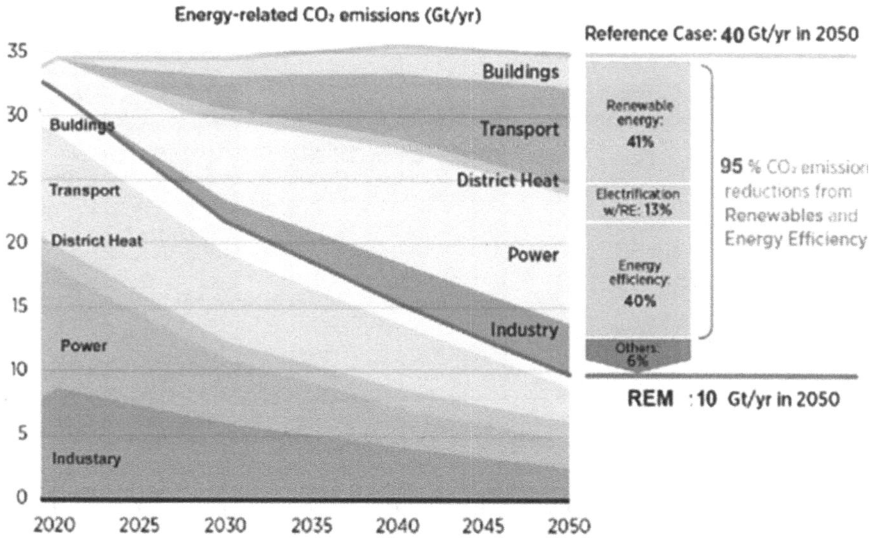

FIGURE 6.5 The possibility of decreasing CO_2 emissions by technology is a comparative analysis of the reference scenario vs the REmap scenario (2020–2050) (Hassan et al., 2024).

Predictions, including the 2050 reference, emphasise the pressing condition of the GHGs. If the present direction is relentlessly followed without considerable shifts, CO_2 discharges from energy-connected sources will rise by 6%, increasing from 32 Gt in 2020 to 37 Gt by 2050, as demonstrated in Figure 6.5. This path differs from the needed yearly reduction of 2.5% in energy-connected CO_2 discharges to align with global goals designed to moderate the worst effects of climate change (Jiang et al., 2023; Liu et al., 2023). The direction worsens global warming and hinders efforts to attain cleaner air in urban areas, unfavourably impacting millions.

To undertake a direction towards sustainability, the energy change must emphasise a significant decrease in emissions. This comprises swift installation of renewable energy resources, enhanced energy efficiency across several areas, and powering innovations in green technologies. Wind and solar energy prices have meaningfully reduced, making them more viable than ever. Joined with developments in energy storage, these renewable sources have the prospect of substituting a large part of fossil-fuel-based energy facilities.

In the renewable energy map (REmap) situation, the key player of innovative energy change develops as support for renewables and sharp energy efficiency is improved. As described in Figure 6.6, this relationship enjoys an unparalleled skill: the possibility to attain an outstanding 95% decrease in emissions by 2050, in an unambiguous disparity to estimates in the reference condition.

This huge decrease emphasises the critical work of renewables and efficiency and measures their individual effects. The synergy between efficiency measures and renewable energy sources does not simply complement; it controls the landscape of possible solutions. While energy efficiency, frequently addressed as the basis of sustainable energy practices, generates an equilibrium by decreasing unnecessary usage and waste, renewables grow ahead to meet energy demands with negligible carbon footprints.

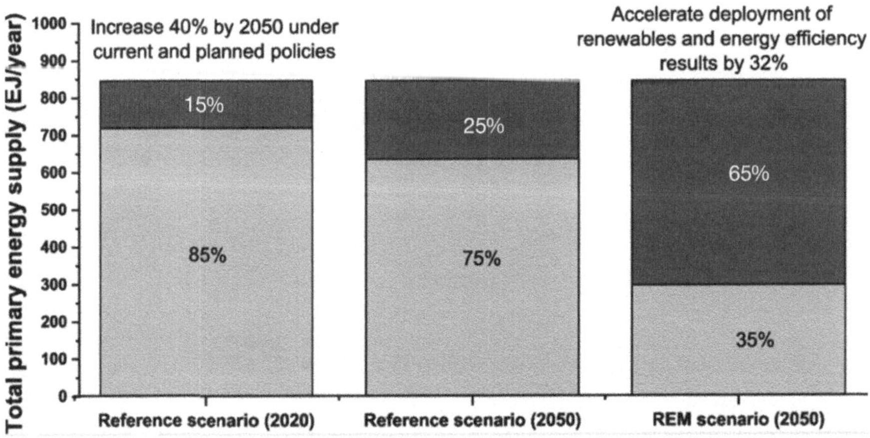

FIGURE 6.6 Global primary energy supply analysis from 2015 to 2050: A comparative analysis of reference scenario vs. REmap scenario.

This coordination indicates that by 2050 renewables could directly subsidise decreasing emissions by a remarkable extent, ranging from 41% to a huge 55%, as depicted in Figure 6.6. This decrease targets echo the hastening global drive behind wind, solar, and other renewable technologies, motivated by their decreasing prices and improved scalability. At the same time, efficiency measures across homes, transportation, and industries highlight the importance of achieving more with less, optimising consumption tactics and incorporating smart technologies. These two initiatives characterise the most effective approach for a sustainable, cleaner future, reforming energy generation and consumption methods.

The drive towards a sustainable energy potential places significant emphasis on the power sector's essential role, which is estimated to make an incredible contribution of more than 15 Gt to the estimated 30 Gt decrease in emissions by 2050. These transformative shifts highlight the significant potential of transformation towards cleaner energy sources within the power sector, surpassing its impact beyond simple electricity generation.

Although the power sector is crucial in energy transformation, other areas equally appear to be significant promoters of this transmutative journey. With their energy demands for cooling, electricity, and heating, buildings can take advantage of developments in energy-efficient strategies and technologies. The industrial area includes manufacturing and processing units with huge prospects for incorporating renewables, optimising energy consumption, and implementing advanced low-carbon technologies. The transport sector, experiencing swift electrification and implementing alternative fuels, can meaningfully decrease its conventionally huge emissions footprint. Moreover, sectors including water treatment, waste management, and agriculture have positive impacts, mostly when approaches like regional heating are prolonged and optimised. The energy transition is vital for distributing social, environmental, and economic significance. Therefore, it is important for the private sector, governments, and financial organisations to immediately reassess their goals, policies, and application strategies to readjust the energy transition with its proposed pathways as presented in Table 6.1.

TABLE 6.1
Short-Term Procedures to Address the Energy Disaster and Hasten the Energy Transition

Ambition	• Growth in the ambition of national renewable energy goals in accordance with climate targets, improved energy security and enhanced affordability. • Set determination of renewable and energy efficiency goals for every end use (transport, electricity, heating and cooling). • Build efficient adoption strategies for every goal.
Institutions	• Make institutions appropriate for the change (for instance, an updated statute for managers, new ministerial structures, and cross-ministry task forces). • Restructure the current loaning services of growth finance organisations by offering more concessional loans and endowments, mainly for countries that experience assets and may be in debt distress.
Physical infrastructure	• Perform combined cross-sector infrastructure development for the energy change with aiming for enlargement (for example, heat networks, power grids, electric vehicle (EV) charging infrastructure, all connected up to improve flexible renewable electricity). • Offer inducements for infrastructure reserves where market obstructions occur (for instance, heat networks, EV chargers). • Rationalise authorising measures for large-scale infrastructure without negotiating environmental and social effect evaluations and ensure public acceptance is adopted. Careers and skills Finance • Established requirements or compulsory goals for new buildings (for example, the number of EV chargers per inhabitant, which links to heat networks). • Offer more public finance to grow the infrastructure needed (for example, through direct possession of resources, such as transmission lines).
Jobs and skills	• Incorporate renewable energy into educational programs; increase vocational education and technical and training prospects. • Accelerate efforts to project future professional requirements in each renewable energy sector and work with industry relations and training institutions to align their arrangement. • Ensure improved access to training possibilities for women, minorities and youth. • Establish pathways for fossil fuel industry workers to reskill and recertify for vocations in renewable energy. This will need public funding for training.
Finance	• Increase and channel public financing – including through international collaboration via a broad spectrum of policies, covering all segments of the renewable energy value chain, the wider energy sector and the economy as a whole. • Strategically plan, select and implement instruments to channel public finance (domestic and international), including (1) government spending such as grants, rebates and subsidies; (2) debt, including existing and new issuances, credit instruments, concessional/blended financing and guarantees; (3) equity and direct ownership of assets (such as transmission lines or land to build projects). • Define 'risk' more comprehensively, going beyond the narrow investor-centric definition of risk (for example, investment in energy assets not paying off) to include broader environmental and social risks. • Continue to use public policy and finance to crowd in private capital.

(*Continued*)

TABLE 6.1 (CONTINUED)
Short-Term Procedures to Address the Energy Disaster and Hasten the Energy Transition

Power sector	• Adopt a power system structure conducive to high shares of variable renewable energy, recognising their techno-economic characteristics. This could include dual energy procurement with long-term procurement through auctions and a short-term flexibility market. • Streamline permitting procedures for renewable power projects without compromising environmental and social impact assessment. Ensure public acceptance is fostered. • Better synchronise power grid expansion and other infrastructure developments with renewable power deployment to avoid bottlenecks. • Design renewable energy procurement processes (e.g. auctions) to serve objectives beyond the lowest price (e.g. development of local industry) and consider design elements to distribute the risks of supply chain disruptions among stakeholders (e.g. indexation of components). • Design policies for self-consumption in a progressive way that supports equitable access to support for the deployment of solutions and the distribution of socio-economic benefits.
End-use sectors – buildings, industry, transport	• Develop energy efficiency programs and measures such as standards in transport, industry, and buildings. Increase finance for energy efficiency through efforts to aggregate projects and de-risk investment. • Promote readiness for new fuels and electrification (e.g. EV chargers; see also Physical Infrastructure). • Behavior changes: incentivise slower driving using speed limits, mandate room temperature limits (e.g. offices and public buildings), reduce individual car usage by promoting public transport and car-sharing, and prefer high-speed and night trains to aircraft where possible.
Cross-sector and crosscutting policies	• Introduce fiscal policy measures: obligations for reinvesting windfall profits of fossil fuel energy revenues in energy transition technologies, reduced subsidies for fossil fuels and raised/newly introduced CO2 prices when fossil fuel prices fall. Ensure the socio-economic benefits of such instruments are distributed fairly. Reform taxes and levies on heating fuels, VAT exemptions for renewables, etc. • Develop national bioenergy and hydrogen strategies (including sectoral prioritisation) to ensure bioenergy and hydrogen can play the most appropriate role in decarbonisation. • Incentivise/mandate a circular economy approach (reduce, reuse, recycle), for example, for energy-intensive products like steel, renewable energy technologies, batteries, cars, etc. This will both reduce energy demand and the demand for critical materials. • Enhance international collaboration across relevant areas, including sustainability governance, energy and climate finance, technology and innovation, regional power grids, and green hydrogen development. • Put greater focus (including through international collaboration) on achieving the universal access targets of SDG7.

As presented in Table 6.1, short-term measures to solve the energy crises and accelerate the energy change require fast actions and plans designed to mitigate energy disasters while hastening the transition from fossil fuels to sustainable energy sources. These procedures characteristically comprise improving energy efficiency, installing renewable energy technologies at scale, stabilising energy supply chains, encouraging demand-side management, and employing policy inducements that support fast decarbonization. The emphasis is on crucial, actionable steps that decrease vulnerability to energy shocks and lay the groundwork for a long-term, sustainable energy change.

Globally, the G20 countries develop as formidable leaders in this transformation, jointly in charge of a surprising 85% renewable installation. A more rapid analysis discloses that China stands out with a contribution of 27% share, indicating its devotion and ability as a renewable energy capital. The United States follows this closely, accounting for a 16% share, reflecting technological developments and plan changes. With its huge population and growing energy demands, India accounts for 15% of wind and solar energy incorporation steps. Similarly, the EU, an alliance with various energy settings, shares 10% of the total, motivated by an integrated green agenda.

Similarly, Canada, which harnesses solar, hydro, and wind prospects, shares 3% of this mix. Other G20 nations, such as Brazil with its bioenergy competence, Russia with huge hydropower and geothermal prospects, and Australia with wind and solar proficiencies, equally exert powerful roles in determining the renewable path. Renewable energy concentration creativities within a restricted number of key players emphasise the serious accountability and influence these countries hold, as their choices, assets, and strategies not only form their specific energy settings but also meaningfully direct the global energy transition description.

6.5 CREDITABLE PRACTICES: TRANSFORMING THE FOOD-ENERGY-WATER-ECOSYSTEMS NEXUS

The practical application of the FEWE nexus produces concrete value as it changes from the theoretical step. The private and public sectors dynamically use the nexus to develop innovative solutions and promote economic growth and business development in riparian countries. Water scarcity is a serious blockage for energy and agriculture, bringing substantial consideration within the nexus agenda. A crucial emphasis is on improving the sustainable consumption of renewable energy, creating an essential part of the innovation procedure. Figure 6.7 summarises vital deliberations for evaluating advantages and prospective trade-offs of renewable energy within a local framework from a nexus perception. Several instances demonstrate the various implementations of the nexus in installing renewable energy (Chen et al., 2023; Zhang et al., 2023).

The first illustration describes utilities' work, demonstrated by Electricité de France (EDF), in partnership with farmers to enhance water distribution across their hydropower plant (Laurenceau, Molle and Grau, 2020). The innovative Water Saving Convention signed by EDF and irrigators led to decreased water usage, benefiting both parties and increasing electricity generation capacities. In the second instance, regionalised renewable energy technologies have developed to increase income sources and reduce energy charges for farms in England and Italy. While uncertainties emerge from variations in feed-in tariffs, improvements in energy storage

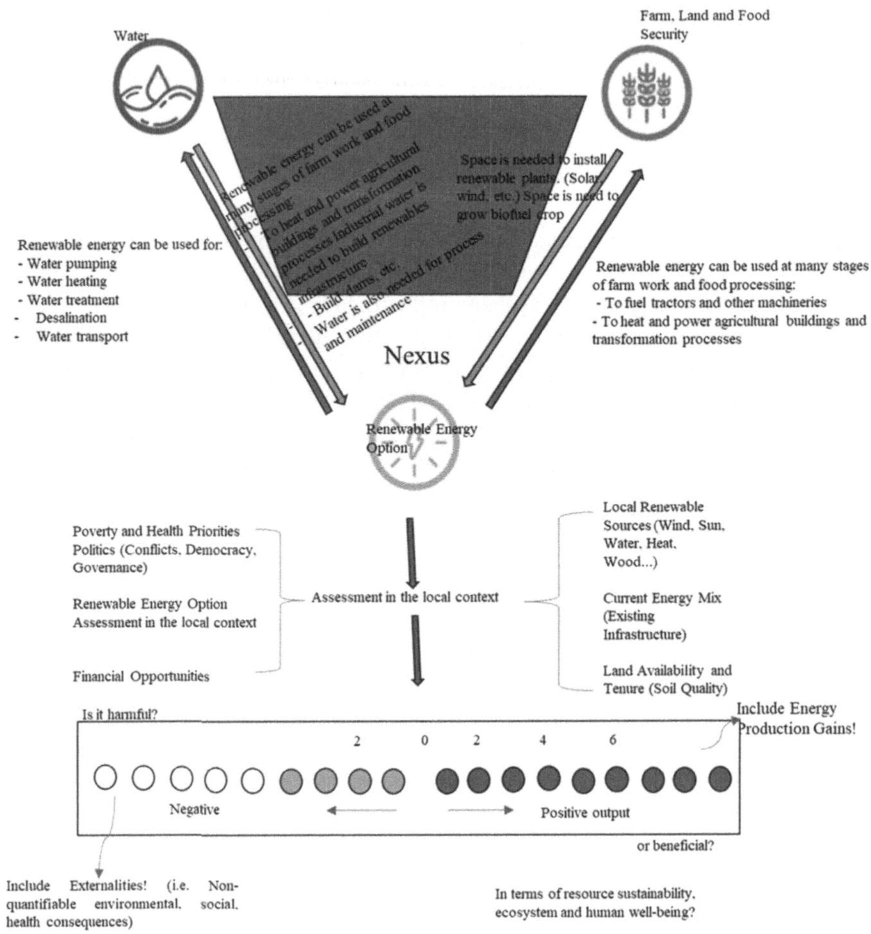

FIGURE 6.7 Developing land, water, and renewable energy trade-offs in establishing renewable energy sources.

hold promise for the prospect of on-farm renewable energy applications. The final illustration relates to controlling bioenergy in the framework of wastewater-treatment amenities, which allows workers to terminate their energy cycle in Germany.

Example 6.1

How utilities could collaborate with farmers to strengthen nexus opportunities: Electricité de France (EDF) (Sarni, 2015): Utilities can substantially exploit nexus prospects, as established by EDF. EDF effectively achieved vital synergies within the FEW nexus by affiliating with key irrigators in the neighbourhood of its Serre-Ponçon hydropower plant. This affiliation led to a contract on the distribution of water resources between power irrigation and generation, encouraging the sustainability of renewable energy applications and harvesting sustenance from all investors concerned. The Serre-Ponçon Dam and reservoir in the Durance and Verdon River system in southeast France consist of 21 hydropower plants generating 6,500 GWh per year of renewable electricity. This system provides drinking and industrial water to the whole area and supplies irrigation water to over 150,000 hectares of farmland. With a definite storage capacity of 450 million cubic metres of water in the summer, the reservoir permits yearly extraction of roughly 1,800 million cubic metres.

In the early 2000s, EDF understood the necessity of discussing water usage right away with irrigators to stop overusage and the possible limitations on power supply during peak demand. Thus, EDF established and implemented a Water Saving Convention, a contract with two key irrigators to close the dam. This contract drew the irrigators' promise to decrease water usage, and EDF responded by reimbursing them for their savings. Significantly, the irrigators had the flexibility to revise their commitments every year. EDF's groundbreaking method of esteeming water included connecting the water price to the energy value it could generate per cubic metre, reproducing present and future energy charges in France. This inducement inspired irrigators to decrease water usage, bringing about technological and process advancements for improved water efficiency. The growing effect was a decrease in yearly water usage from 325 million cubic metres between 2000 and 2006 to 235 million cubic metres in 2015. On the side of EDF, these savings are interpreted as a bigger electricity generation capacity during peak demand periods. In 2014, a subsequent contract building upon the innovative Water Saving Convention reiterated the value formed for all concerned parties.

Example 6.2

How renewable energy explains nexus relationships at the farm level (Todman, 2017). The applications of on-farm renewable technology expand income sources, reduce energy prices, and improve flexibility. The energy generated can be sold to the main grid or used on the farm, depending on supervisory support. In England, 23% of farms participated in renewable energy generation in 2015, sharing roughly 9% of the entire income (Vanhuyse, 2016; Kallio et al., 2020). While uncertainties restricting fluctuations in feed-in tariffs prevent more power generation deployments on English farms, improvements in energy storage could produce possible solutions. Italy experienced the same situation because of an effective feed-in tariff system; 13% of the deployed solar PV capacity was connected to the agricultural sector, bringing out fears of speculation and rivalry with farms' essential agricultural activity (Sawin, 2012).

However, terminating the scheme and favouring self-consumption and small installations are expected to mitigate these risks. Meanwhile, commitment to energy generation and consumption in agriculture is obvious in Italy, with joint efforts by the Ministry of Energy and Forestry and the Italian National Agency for New Technologies, Energy, and Sustainable Economic Development (Magnani and Carrosio, 2021). This experience highlights the significance of flexible policy regulations as circumstances warrant. Renewable energy's influence on the nexus is complicated, manipulating water quality positively through activities such as the anaerobic digestion of farmyard manure. Nevertheless, negative effects, such as erosion during construction and improved crop farming for digestion, must be considered. Effects on food production and ecosystems, such as shifts in microclimate and dangers in habitat, including bird collisions and fish migration with wind turbines, are serious concerns. The nexus method should be controlled to attain higher sustainable development of renewable energy sources. Estimating in advance how shifts in energy production could influence the water, food sectors, or ecosystems, in general, may prevent unplanned costs and capitalise on win-wins between the sectors.

Example 6.3

How the prospect of energy in wastewater addresses urban nexus: The Jenfelder Au eco-district's innovative method to urban nexus connections by addressing water–energy linkages through bioenergy opportunities. As an aspect of the Hamburg Water Cycle, this closed-loop system enhances resource consumption by incorporating wastewater treatment and energy production. Wastewater is collected, channelled to a biogas plant, and used for electricity and heat generation for a region with 770 units and 2,000 inhabitants. Grey water is recycled for flushing and irrigation, while rainwater is incorporated for extra irrigation uses. Hamburg Wasser, the municipal water company, supported the scheme, initially indicating its viability in an environmental theme park, Gut Karlshöhe, where it was activated for educational aims before its implementation by the Wandsbek District Authority and the Jenfelder Au district. The system provides various advantages, such as decreased external energy demands in wastewater treatment, stress decrease on stormwater infrastructure, and zero CO_2 emissions. The disintegration of black and grey water and the advancement of onsite green areas lower the stress on stormwater facilities. This moderates flooding risks and improves flexibility to climate change, with possible agricultural applications through recycling biogas plant fermentation remains.

6.6 FEW NEXUS IN SUSTAINABLE DEVELOPMENT AND ENVIRONMENTAL MANAGEMENT

The FEW nexus approach implements a pragmatic method to simplify the balanced, efficient, and logical management of the environment, socio-economic systems, and natural resources (Flammini et al., 2014). With a complete and multi-sectoral technique, the FEW nexus agenda encourages sustainable development and addresses the challenges posed by global environmental change (Chang et al., 2016; Miralles-Wilhelm, 2016). Consequently, efficient management of the nexus is important for attaining the UN SDGs (Giupponi and Gain, 2017), as it tackles the serious features of food, energy, and water security, along with environmental management, through global institutional partnership and financial inducements (Mohtar, 2016). Human-induced distractions and modern concentrated agriculture have significantly stressed FEW resources, delaying improvement towards SDGs and environmental management (De Amorim et al., 2018). The inception of the COVID-19 outbreak has also worsened the challenges in accomplishing SDGs, demanding intensive consideration in future studies (Cheng et al., 2021).

Hence, it is vital during the epidemic to improve institutional capacity and increase awareness about the limitations connected with the FEW nexus (Martinez-Hernandez, Leach and Yang, 2017). Earlier research efforts have tried to prove the utility of the nexus for combined sustainable development and environmental management (Menegaki and Tiwari, 2018). Griggs has also suggested a more articulate social and environmental agenda with measurable targets regarding the SDGs (Naidoo and Fisher, 2020). The elements of the FEW nexus are complicatedly connected to definite SDGs, with the prospect of decreasing environmental effects (Menegaki and Tiwari, 2018). Besides SDGs, Figures 6.8 and 6.9 explain how the FEW nexus offers a wider agenda for advancing the Circular Economy, including better-quality design and materials management (Wang et al., 2021). The incorporation of the FEW nexus into environmental management strategies can meaningfully contribute to UN-SDGs, circular economy enterprises, and climate action, all of which are crucially connected with sustainable development (Liu et al., 2018; Hoosain et al., 2023).

The innovation of the nexus strategy starts with its practical engagement to tackle critical challenges, including market access, equity, and land tenure, which have historically been ignored, notwithstanding their complex relationships with environmental management (Blaikie, 2016). The FEW nexus ideas propose a multipurpose agenda appropriate across several industries to develop environmental sustainability.

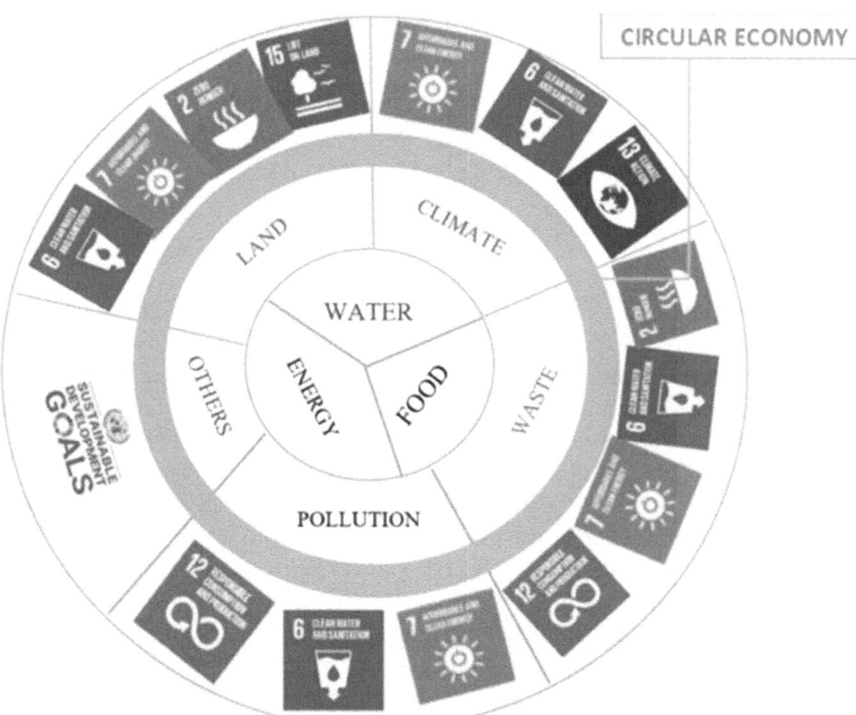

FIGURE 6.8 Connecting the FEW nexus with the circular economy and SDGs (Rai, 2023).

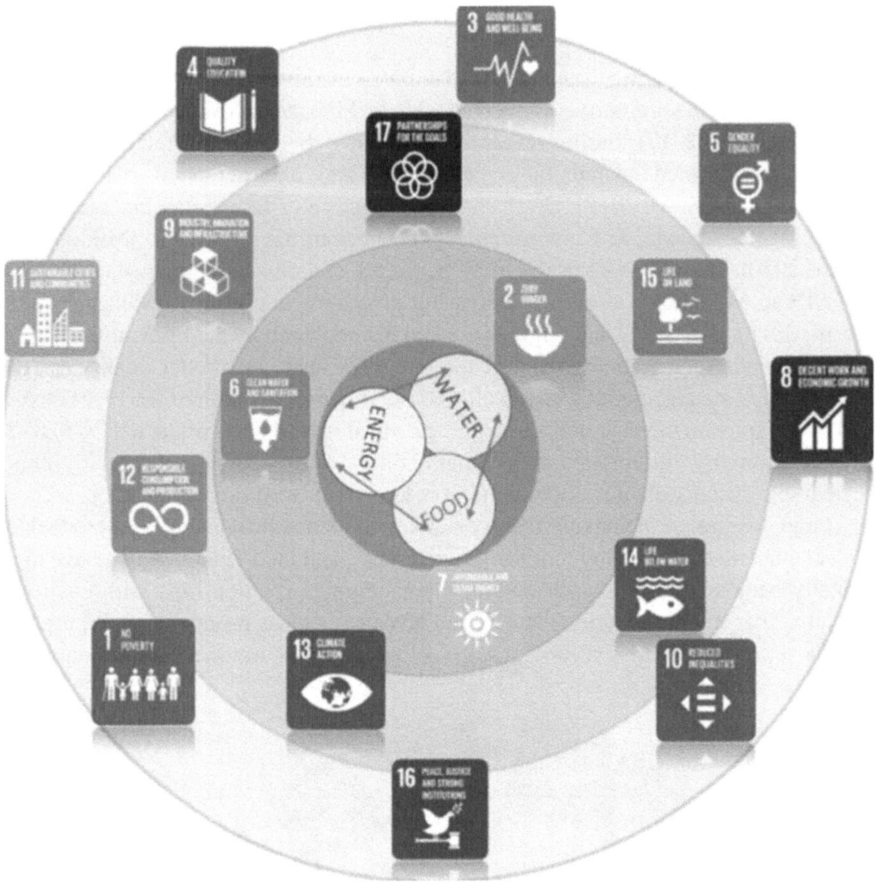

FIGURE 6.9 Influences of nexus strategies on SDGs (Rai, 2023).

For instance, farm efficiency can be accomplished by implementing sustainable activities such as biogas production, rainwater harvesting, bio-fertiliser utilisation, and solar energy generation (de Castro Sobrosa Neto et al., 2018). These approaches alleviate reliance on non-renewable energy sources and suggest viable pathways towards attaining FEW security (de Castro Sobrosa Neto et al., 2018).

Stochastic models and probabilistic analysis are needed to encourage accurate nexus data to assist in risk management decision-making (Karan et al., 2018). Integrating stochastic elements can alleviate global risks and foster improved system integration (Bogardi et al., 2012). The FEW nexus helps decision-makers from an "inside-out viewpoint," bearing in mind myriad interrelated challenges (Rai, 2023). Effective governance can stimulate the FEW nexus and facilitate sustainable development alongside environmental stewardship (Bogardi et al., 2012).

Encouraging financial inducements to drive technological invention can accelerate growth towards energy flexibility and SDGs within the FEW basis (de Andrade Guerra et al., 2021). Increasing the analysis, collection, sharing, and discussion of multi-sectoral FEW nexus data improves knowledge modification and offers empirical support for informed environmental management practices (Martinez-Hernandez, Leach and Yang, 2017). Therefore, cross-sectoral FEW connections allow sustainable environmental management by clearly addressing trade-offs and leveraging synergies (Urbinatti et al., 2020).

6.7 ADDRESSING ENERGY RESILIENCE AND SDGS IN THE COVID-19 OUTBREAK

6.7.1 Energy Resilience

The energy sector and its ecological footprint have experienced several changes due to the COVID-19 outbreak (Jiang et al., 2021). Shutdowns across businesses and decreased transportation activities originally led to a worldwide drop in energy usage (Klemeš et al., 2020). Nevertheless, continued periods of staying at home led to temporary changes in energy demand patterns, causing more stress on electrical grids (Nundy et al., 2021). Extended lockdowns gave rise to dependence on energy-intensive devices required for remote work and education, thereby increasing an urgent demand for energy-efficient buildings to withstand development towards environmental sustainability (Nundy et al., 2021).

The advancement of renewable energy resources can promote both environmental sustainability and reduce the spread of the COVID-19 pandemic (Coccia, 2020). However, the outbreak has brought challenges to the renewable energy industry, with bigger plastic and solid waste production delaying its growth (Jiang et al., 2021). Disposable facemasks, for example, were allegedly observed to obstruct municipal solid waste anaerobic digestion, causing a decrease in methane generation by up to 18% (de Albuquerque et al., 2021). Besides, interruptions in transportation have affected the circulation of renewable energy resources to areas, mainly in developing countries that depend on weak biomass sources such as wood-fired cooking (Ravindra et al., 2021). Also, the COVID-19 pandemic has interrupted grants intended to facilitate renewable energy implementation (Tsao et al., 2021). Attaining the UN SDGs demands incorporating several initiatives, including increasing access to renewable energy in local sectors, encouraging water conservation, and executing carbon pricing taxes (Barbier and Burgess, 2020). Programmes such as installing renewable energy systems, expanding biorefineries, and preserving water resources can contribute to the development of sustainable rural development and a circular economy.

Furthermore, solidifying resilient components within the energy sector is essential to tackling socio-economic and health challenges and research in climate action (Jiang et al., 2021). Resilient energy systems are flexible for investors, targeting change towards carbon-neutral economies and the low-carbon aftermath of COVID-19 (Heffron et al., 2021). Despite COVID-19 being a major interrupter, it has equally offered chances to discover low-impact alternatives to traditional power generation approaches (Kanda and Kivimaa, 2020). Predicting energy sustainability within the outbreak can assist as a roadmap towards wider environmental sustainability targets (Elavarasan et al., 2021).

6.7.2 Sustainable Development Goals

The UN's SDGs, established in 2015, are designed to assist in attaining global public health and improve human welfare and environmental sustainability by 2030. However, the inception of COVID-19 hindered the timely achievement of 17 of these SDGs, meaningfully affecting ecosystem flexibility, human health, and livelihoods (Rai et al., 2022). The prevalent impacts of COVID-19 on the global community, ecosystems, and economies have posed difficult challenges to the timely accomplishment of the SDGs (Barbier and Burgess, 2020). Throughout the epidemic, efforts to prioritise safety and health, mostly in connection with SDG 3, took priority, while the remaining 16 SDGs were considered less important. Concerns expressed by the editors of *The Lancet Public Health* (Health, 2020) emphasised reduced initiatives to ease poverty, with predictions showing that 6% of the worldwide population may still undergo extreme poverty by 2030. The COVID-19 disaster has pushed 71 million people into severe poverty, signifying a huge hindrance to SDG 1 (poverty reduction). Economic downturns caused by the outbreak have worsened disparities in food security, healthcare, and means to clean water globally, strictly undermining SDGs 2, 3, 6, and 10.

Additionally, the search for SDG 4 (complete and reasonable access to education) has been significantly delayed by COVID-19, mostly in regions with insufficient telecommunication facilities (Barbier and Burgess, 2020). Similarly, lockdown measures during the outbreak have equally destabilised growth towards SDG 5 (gender equality), with a bigger workload imposed on females causing a 30% rise in worldwide reports of domestic violence (Berchin and de Andrade, 2020). To resolve the complex effects of COVID-19, a redescription of the sustainability pattern in livelihoods, governance, and economy is important. Besides, the interruption of SDGs 16 (promoting peace and safety from violence) and 17 (solidifying global collaborations) during the outbreak (Berchin and de Andrade, 2020) has been ascribed to global unrest surrounding the control of COVID-19's cause and spread, socioeconomic differences, injustices in healthcare, and the failure to ensure reasonable access to crucial medicines.

The outbreak's impact on bioeconomies and energy resilience are strongly connected with SDGs (Woźniak and Tyczewska, 2021). Unsatisfactory progress towards SDG 1 has forced local communities to opt for biomass-based fuels for domestic cooking, impeding the accomplishment of SDG 7 (clean energy) and SDG 13 (climate action). This change to biomass within COVID-19 worsened air pollution among poor local communities (Mikulčić, Baleta and Klemeš, 2022), posing health dangers and hindering the development of SDGs 1–5, 7–8, 11, 13, and 15. On the other hand, changing from biomass to wind turbines and solar arrays could accelerate SDG 7 progress (Elavarasan et al., 2021). Stressing a preference for renewable bioresource plans for improving circular bio economies and energy resilience during the outbreak is necessary (Mikulčić, Baleta and Klemeš, 2022).

The impacts of COVID-19 require evaluation from economic, social, political, ecological, and environmental viewpoints (Patterson et al., 2021). The outbreak has reformed plans, ideas, and the search for environmental sustainability from a long-standing viewpoint (Guo and Lee, 2022). At the same time, massive vaccination campaigns must verify sustainability parameters and life-cycle (Mikulčić, Baleta and Klemeš, 2022). A recent life-cycle analysis of COVID-19 vaccination plans based on the 4Es (economic, equity, energy, and environmental) discloses their unsustainability in addressing the SDG accomplishment (Jiang et al., 2021). The 4E life-cycle evaluation of COVID-19 vaccination initiatives showed that such energy-intensive footprints would negatively affect SDG 7. Incorporating global institutional efforts is necessary to protect sustainable COVID-19 vaccination initiatives (Jiang et al., 2021; Mikulčić, Baleta and Klemeš, 2022).

Climate activities are complicatedly tied to SDGs (Klemeš et al., 2020) as with energy technology and policy. Nevertheless, the recent climate action regulatory plan, including the Paris Agreement and SDG 13, experienced negotiations because of COVID-19 (Wyns and van Daalen, 2021), affecting SDG 1 through 56 sustainability measures (Soergel et al., 2021). Submissions incorporating the SDG timeframe with a "pandemic rearrange" to 2050 have appeared (Soergel et al., 2021). Yet, some studies suggest that COVID-19 has highlighted difficulties in SDG accomplishment, encouraging targeted solutions instead of reviewing timeframes (Ottersen and Engebretsen, 2020). In essence, while the outbreak worsened healthcare challenges and environmental intimidations (Rai et al., 2022), it equally recognised gaps in worldwide environmental strategies, encouraging the implementation of sustainable procedures across areas such as climate action research, rural livelihoods, energy, and human well-being.

6.8 CONCLUSION

Renewable energy could play a vital role in supporting better management of resources and encouraging synergies within the nexus of FEWE. Implementing a nexus strategy is in accordance with the sustainable development programme for 2030, enabling faster partnerships across areas and countries and developing informed conversations on trade-offs. Incorporating the nexus strategy into the SDG agenda offers an opportunity to assist industries and decision-makers with mutual targets to encourage sustainable development while connecting renewable energy resources efficiently. Renewable energy contributes to solving nexus challenges, especially in streamlining transboundary collaboration. Conducting nexus analysis can improve the sustainability of hydropower and bioenergy programmes by offering broader visions of their influences on land, water, and ecosystems. By integrating synergies and effects into arrangement and strategy development, wider utilisation and acceptance of renewable energy, including hydropower, can be accomplished, even when trade-offs are more noticeable. Climate change is a buttressing factor for the nexus method, highlighting the necessity of discourse adaptation and mitigation approaches. With the pressures climate change brings to energy, water, and land resources, implementing a nexus strategy highlights possibilities for mitigating the challenges. Renewable energy offers several measures to combat the mitigations, including decreasing traditional energy utilisation, reducing CO_2 emissions, improving resource productivity, and conserving ecosystems. Decision-makers could have several tools to examine nexus opportunities for renewable energy, such as policy guidelines and national environmental standards. Integrating the nexus plan into industry performance presents practical advantages, including better risk management and resource efficiency. While executing macro-level changes may be challenging, streamlining organisational processes at the local level can enable development. Transboundary collaborations are crucial for accomplishing concrete results from the nexus strategy, especially regarding sustainable renewable energy growth.

REFERENCES

Al-Saidi, M. and Elagib, N. A. (2017) 'Towards understanding the integrative approach of the water, energy and food nexus'. *Science of the Total Environment*, 574, pp. 1131–1139.

Apeh, O. O., & Nwulu, N. (2024a). The food-energy-water nexus optimization: A systematic literature review. *Research on World Agricultural Economy*, 5(4).

Apeh, O. O., & Nwulu, N. I. (2024b). The water-energy-food-ecosystem nexus scenario in Africa: Perspective and policy implementations. *Energy Reports*, 11, 5947–5962.

Apeh, O. O., & Nwulu, N. I. (2025). Machine learning predictions for fault detections in solar photovoltaic systems: A bibliographic outlook. *Journal of Infrastructure, Policy and Development*, 9(2), Article 9940.

Atlantic Council. (2021) *The global energy agenda. Report*, Washington, DC.

Barbier, E. B. and Burgess, J. C. (2020) 'Sustainability and development after COVID-19'. *World Development*, 135, p. 105082.

Beddington, J. (n.d.) *Food, energy, water and the climate: A perfect storm of global events?* Chief Scientific Adviser to HM Government.

Berchin, I. I. and de Andrade, J. B. S. O. (2020) 'GAIA 3.0: Effects of the coronavirus disease 2019 (COVID-19) outbreak on sustainable development and future perspectives'. *Research in Globalization*, 2, p. 100014.

Blaikie, P. (2016) *The political economy of soil erosion in developing countries*. Routledge.

Bleischwitz, R., Johnson, C. M. and Dozler, M. G. (2014) 'Re-Assessing resource dependency and criticality. Linking future food and water stress with global resource supply vulnerabilities for foresight analysis'. *European Journal of Futures Research*, 2, pp. 1–12.

Bogardi, J. J. et al. (2012) 'Water security for a planet under pressure: interconnected challenges of a changing world call for sustainable solutions'. *Current Opinion in Environmental Sustainability*, 4(1), pp. 35–43.

Chang, Y. et al. (2016) 'Quantifying the water-energy-food nexus: Current status and trends'. *Energies*, 9(2), p. 65.

Chen, H. et al. (2023) 'Nexus between energy poverty and sustainable energy technologies: a roadmap towards environmental sustainability'. *Sustainable Energy Technologies and Assessments*, 56, p. 102949.

Cheng, Y. et al. (2021) 'Global action on SDGs: Policy review and outlook in a post-pandemic era'. *Sustainability*, 13(11), p. 6461.

Chhandama, M. V. L. and Rai, P. K. (2023) 'Coupling bioremediation and biorefinery prospects of microalgae for the circular economy'. *Bioresource Technology Reports*, p. 101479.

Coccia, M. (2020) 'How (un) sustainable environments are related to the diffusion of COVID-19: The relation between coronavirus disease 2019, air pollution, wind resource and energy'. *Sustainability*, 12(22), p. 9709.

de Abreu, M. R. and Machado, R. L. (2023) 'Water-energy-food-land nexus challenges and contributions to the biofuel supply chain: Systematic review and meta-synthesis'. *Environment and Development*, 48, p. 100927.

de Albuquerque, F. P. et al. (2021) 'Fate of disposable face masks in high-solids anaerobic digestion: experimental observations and review of potential environmental implications'. *Case Studies in Chemical and Environmental Engineering*, 3, p. 100082.

De Amorim, W. S. et al. (2018) 'The nexus between water, energy, and food in the context of the global risks: An analysis of the interactions between food, water, and energy security'. *Environmental Impact Assessment Review*, 72, pp. 1–11.

de Andrade Guerra, J. B. S. O. et al. (2021) 'A literature-based study on the water-energy–food nexus for sustainable development'. *Stochastic Environmental Research and Risk Assessment*, 35, pp. 95–116.

de Castro Sobrosa Neto, R.. et al. (2018) 'An integrative approach for the water-energy-food nexus in beef cattle production: A simulation of the proposed model to Brazil'. *Journal of Cleaner Production*, 204, pp. 1108–1123.

Elavarasan, R. M. et al. (2021) 'Envisioning the UN Sustainable Development Goals (SDGs) through the lens of energy sustainability (SDG 7) in the post-COVID-19 world'. *Applied Energy*, 292, p. 116665.

Flammini, A. et al. (2014) *Walking the nexus talk: assessing the water-energy-food nexus in the context of sustainable energy for all initiatives, including environmental and natural resource management*. Working Paper (FAO) eng no. 58.

Gielen, D. et al. (2019) 'The role of renewable energy in the global energy transformation'. *Energy Strategy Reviews*, 24, pp. 38–50.

Giupponi, C. and Gain, A. K. (2017) 'Integrated spatial assessment of the water, energy and food dimensions of the sustainable development goals'. *Regional Environmental Change*, 17, pp. 1881–1893.

Guo, Q. and Lee, D. C. (2022) 'The ecology of COVID-19 and related environmental and sustainability issues'. *Ambio*, 51(4), pp. 1014–1021.

Hassan, Q. et al. (2024) 'The renewable energy role in the global energy transformations'. *Renewable Energy Focus*, 48, p. 100545.

Health, T. L. P. (2020) 'Will the COVID-19 pandemic threaten the SDGs?' *The Lancet. Public Health*, 5(9), p. e460.

Heffron, R. J. et al. (2021) 'The role of flexibility in the light of the COVID-19 pandemic and beyond: Contributing to a sustainable and resilient energy future in Europe'. *Renewable and Sustainable Energy Reviews*, 140, p. 110743.

Hoosain, M. S. et al. (2023) 'The influence of circular economy and 4IR technologies on the climate–water–energy–food Nexus and the SDGs'. *Water*, 15(4), p. 787.

International Renewable Energy Agency (IRENA) (2022) 'World energy transitions outlook 2022'. *World Energy Transitions*, pp. 1–54. https://irena.org/Digital-Report/World-Energy-Transitions-Outlook-2022%0A. https://irena.org/publications/2021/March/World-Energy-Transitions-Outlook

Jiang, J. et al. (2023) 'Low-carbon energy policies benefit climate change mitigation and air pollutant reduction in megacities: An empirical examination of Shenzhen, China'. *Science of the Total Environment*, 892, p. 164644.

Jiang, P. et al. (2021) 'Energy, environmental, economic and social equity (4E) pressures of COVID-19 vaccination mismanagement: A global perspective'. *Energy*, 235, p. 121315.

Kallio, L. et al. (2020) 'Farm power: How a new business model impacts the energy transition in Finland'. *Energy Research & Social Science*, 65, p. 101484.

Kanda, W. and Kivimaa, P. (2020) 'What opportunities could the COVID-19 outbreak offer for sustainability transitions research on electricity and mobility?' *Energy Research & Social Science*, 68, p. 101666.

Karan, E. et al. (2018) 'Towards the optimization of sustainable food-energy-water systems: A stochastic approach'. *Journal of Cleaner Production*, 171, pp. 662–674.

Klemeš, J. J. et al. (2020) 'Minimising the present and future plastic waste, energy and environmental footprints related to COVID-19'. *Renewable and Sustainable Energy Reviews*, 127, p. 109883.

Laurenceau, M., Molle, F. and Grau, M. (2020) 'Reducing water withdrawals: the negotiation and implementation of environmental policy in the Durance River Basin, France'. *Water Policy*, 22(6), pp. 1217–1236.

Liu, J. et al. (2018) 'Nexus approaches to global sustainable development'. *Nature Sustainability*, 1(9), pp. 466–476.

Liu, L. J. et al. (2023) 'Carbon emissions and economic impacts of an EU embargo on Russian fossil fuels'. *Nature Climate Change*, 13(3), pp. 290–296.

Liu, Q. (2016) 'Interlinking climate change with water-energy-food nexus and related ecosystem processes in California case studies'. *Ecological Processes*, 5(1), p. 14.

Mabhaudhi, T. et al. (2016) 'Southern Africa's water-energy nexus: Towards regional integration and development'. *Water*, 8(6), p. 235.

Magnani, N. and Carrosio, G. (2021) *Understanding the energy transition: Civil society, territory and inequality in Italy*. Springer Nature.

Martinez-Hernandez, E., Leach, M. and Yang, A. (2017) 'Understanding water-energy-food and ecosystem interactions using the nexus simulation tool NexSym'. *Applied Energy*, 206, pp. 1009–1021.

Menegaki, A. N. and Tiwari, A. K. (2018) 'A global food–energy–water nexus with heterogeneity, non-stationarity and cross-sectional dependence'. *Quality & Quantity*, 52, pp. 2723–2755.

Mikulčić, H., Baleta, J. and Klemeš, J. J. (2022) 'Cleaner technologies for sustainable development'. *Cleaner Engineering and Technology*, 7, p. 100445.

Miralles-Wilhelm, F. (2016) 'Development and application of integrative modeling tools in support of food-energy-water nexus planning—A research agenda'. *Journal of Environmental Studies and Sciences*, 6, pp. 3–10.

Mohtar, R. (2016) *The importance of the water-energy-food nexus in the implementation of the Sustainable Development Goals (SDGs)*. OCP Policy Center.

Mohtar, R. H. (2022) 'The WEF nexus journey'. *Frontiers in Sustainable Food Systems*, 6, p. 820305.

Muller, M. (2015) 'The 'nexus' as a step back towards a more coherent water resource management paradigm'. *Water Alternatives*, 8(1), 675–694.

Naidoo, R. and Fisher, B. (2020) 'Reset sustainable development goals for a pandemic world'. *Nature*, 583(7815), pp. 198–201.

Niang, A. et al. (2017) 'Variability and determinants of yields in rice production systems of West Africa'. *Field Crops Research*, 207, pp. 1–12.

Nundy, S. et al. (2021) 'Impact of COVID-19 pandemic on socio-economic, energy-environment and transport sector globally and sustainable development goal (SDG)'. *Journal of Cleaner Production*, 312, p. 127705.

Nzuma, J. M. et al. (2010) 'Strategies for adapting to climate change in rural sub-Saharan Africa: A review of data sources, poverty reduction strategy programs (PRSPs) and National Adaptation Plans for Agriculture (NAPAs) in ASARECA member countries'. *IFPRI discussion papers*, 1013. https://hdl.handle.net/10568/33403

Ottersen, O. P. and Engebretsen, E. (2020) 'COVID-19 puts the sustainable development goals center stage'. *Nature Medicine*, 26(11), pp. 1672–1673.

Patterson, J. et al. (2021) 'The political effects of emergency frames in sustainability'. *Nature Sustainability*, 4(10), pp. 841–850.

Polcyn, J. et al. (2023) 'Evaluating the influences of health expenditure, energy consumption, and environmental pollution on life expectancy in Asia'. *International Journal of Environmental Research and Public Health*, 20(5), p. 4000.

Rai, P. K. (2023) 'Role of water-energy-food nexus in environmental management and climate action'. *Energy Nexus*, 11, p. 100230.

Rai, P. K. et al. (2022) 'The effects of COVID-19 transmission on environmental sustainability and human health: Paving the way to ensure its sustainable management'. *Science of the Total Environment*, 838, p. 156039.

Ravindra, K. et al. (2021) 'Impact of the COVID-19 pandemic on clean fuel programmes in India and ensuring sustainability for household energy needs'. *Environment International*, 147, p. 106335.

Sarni, W. (2015) 'Deflecting the scarcity trajectory: Innovation at the water, energy, and food nexus [WWW Document]'. https://dupress.deloitte.com/dup-us-en/deloitte-review/issue-17/waterenergy-food-nexus.html (accessed 5.2.2017).

Sawin, J. L. (2012) 'National policy instruments: Policy lessons for the advancement and diffusion of renewable energy technologies around the world'. In Janet L. Sawin (Ed.), *Renewable energy* (pp. 71–114). Routledge.

Segovia-Hernández, J. G., Contreras-Zarazúa, G. and Ramírez-Márquez, C. (2023) 'Sustainable design of water-energy-food nexus: A literature review'. *RSC Sustainability*, 1(6), 1332–1353.

Soergel, B. et al. (2021) 'A sustainable development pathway for climate action within the UN 2030 Agenda'. *Nature Climate Change*, 11(8), pp. 656–664.

Todman, L. (2017) 'On-farm renewables and resilience: A water-energy-food nexus case study'. In *EGU general assembly conference abstracts* (p. 8070).

Tsao, Y.-C. et al. (2021) 'COVID-19: Government subsidy models for sustainable energy supply with disruption risks'. *Renewable and Sustainable Energy Reviews*, 150, p. 111425.

Urbinatti, A. M. et al. (2020) 'The conceptual basis of water-energy-food nexus governance: Systematic literature review using network and discourse analysis'. *Journal of Integrative Environmental Sciences*, 17(2), pp. 21–43.

Vanhuyse, F. (2016) *The impact of management practices on financial performance: evidence from farm businesses in England*. University of Reading.

Wang, X.-C. et al. (2021) 'Extended water-energy nexus contribution to environmentally-related sustainable development goals'. *Renewable and Sustainable Energy Reviews*, 150, p. 111485.

Waughray, D. (2011) *Water security, the water-food-energy-climate nexus: The world economic forum water initiative*. Island Press.

Woźniak, E. and Tyczewska, A. (2021) 'Bioeconomy during the COVID-19 and perspectives for the post-pandemic world: Example from EU'. *EFB Bioeconomy Journal*, 1, p. 100013.

Wyns, A. and van Daalen, K. R. (2021) 'From pandemic to Paris: the inclusion of COVID-19 response in national climate commitments'. *The Lancet Planetary Health*, 5(5), pp. e256–e258.

Zhang, P. et al. (2023) 'Nexus of institutional quality and technological innovation on renewable energy development: The moderating role of green finance'. *Renewable Energy*, 214, pp. 233–241.

7 Challenges and Opportunities Facing Humans and Natural Resources in the 21st Century

7.1 INTRODUCTION

In the contemporary world, anthropogenic climate change threatens human and environmental systems. These dangers are gradually apparent daily through a stream of climate-connected threats. The future footprints of the planet, including well-being, developmental prospects, and ecosystems, are a rising challenge. Policymakers are examining these threats and exploring possible strategies to mitigate them. Across the world, especially among younger generations, there is a huge demand for decisive, determined action. Companies struggle to navigate emerging risks to their economic stability while also exploring opportunities for growth. Besides, within the increasing anxiety about climate change, the COVID-19 pandemic has disclosed susceptibilities, intensified climate-connected threats, and encouraged reflections on inspiring a sustainable, resilient, and comprehensive recovery.

Although extensive efforts have been advocated recently, anthropological advancement still faces limitations. Almost one-seventh of the global population faces food insecurity and restricted access to crucial resources, including clean water, healthcare, and modern energy sources (Mohtar and Daher, 2016). This crisis considers food shortages, energy limitations, and water scarcity as the main threats from a global viewpoint. Furthermore, the increasing population aggravates the demand for critical resources such as food, energy, and water. Several initiatives have been projected to address human needs and exploit available resources (Liu and Chen, 2020; Cansino-Loeza and Ponce-Ortega, 2021; Hua et al., 2021). For example, the Twin Track and Integrated Water Resources Management (IWRM) strategies are proposed for food and water subsystem management (Gain, Rouillard and Benson, 2013). However, these one-dimensional models concentrating on specific subsystems are unable to explain the complex connections among resources (Cansino-Loeza and Ponce-Ortega, 2021), despite their integral interrelationship (Leck et al., 2015). Indeed, actions performed within one area always affect other sectors (Molajou et al., 2021); hence, performing such one-dimensional methods is insufficient for managing resources with intricate linkages (Hoff, 2011; Ma et al., 2021). For instance, while

DOI: 10.1201/9781003633846-7

IWRM concentrates on water sustainability, it may unintentionally abandon the sustainability of the food and energy sectors. Definitely, policies and decisions exclusively concentrate on one subsystem to address sustainability and security threats, overlooking the demands of other interrelated subsystems (Apeh and Nwulu, 2024).

7.2 IDENTIFICATION OF MAJOR CHALLENGES ASSOCIATED WITH REGIONS AND NATURAL RESOURCES

Climate change offers a broad diversity of hazards across sectors, regions, and scales, which could deteriorate meaningfully under definite risks, exposure, and vulnerability conditions. While some of these threats may happen, others might not. At a universal measure, major risks include threats to biodiversity across coastlines, oceans, and terrestrial ecosystems, mostly in biodiversity hotspots. The study of vulnerability evaluation results of the INFORM Risk Index and WorldRiskIndex at the regional level in combination with population data demonstrates a high concentration of highly vulnerable areas. It illustrates that the global nexus of human vulnerability is not only in regions but also frequently appears within regional groups, mostly in Africa, Asia, and Central America, as shown in Figure 7.1.

A collected regional summary of certain indicators utilised within the vulnerability index is mapped in Figure 7.1. The summary indicates that several combined challenges experienced by African regions are harshly obvious, but equally in other areas, particularly Asia and Central and South America. There are many challenges, including dislocation, inequality, and governance issues, which entirely give rise to vulnerability and limit the adaptive capacities of these areas to climate change.

These threats comprise ecological disorders, including changes in food patterns, energy, tree mortality, population decrease, and decrease in growth, particularly affecting shellfish. These environmental challenges have cascading impacts on incomes and food safety. Furthermore, global-scale threats spread across property, people, and facilities, demonstrating phenomena such as river flooding, severe heat (mainly in urban locations), and threats to fisheries that affect living standards and food safety, as well as specific health threats such as psychopathologies and food-borne diseases. Across sectors and countries, economic limitations are acknowledged as essential and resulting in limits to adaptation, mostly in low- to middle-income regions, as illustrated in Table 7.1, while signs of climate change influencing economic resource availability are presented in Table 7.2.

Major risks around definite systems, regions, or demographic groups show sharp importance. Small islands and coastal regions serve as hotspots for various risks, influencing ecosystems (particularly coral reefs), infrastructure, human health, and incomes. Polar regions experience socio-ecological threats, while South America's African savannahs and Amazon Rainforest are challenged with ecological risks. Wildfire hazards mainly affect North America and Australasia, whereas vector-borne diseases pose difficulties in Asia and Africa. Besides, depriving cultural heritage is a risk in many areas such as Africa, North America, islands, Australasia, and the mountainous regions. Low-income inhabitants are mostly exposed to major risks. Climate-associated effects intensify starvation and food uncertainty, excessively influencing

FIGURE 7.1 Global vulnerability map focusing on the INFORM risk index and WorldRiskIndex (Birkmann et al., 2022).

TABLE 7.1
Main Limitations Related to Restrictions to Adaptation for Regions

Regions	Main Limitations Related to Natural Resource Adaptation
Africa	Economic constrictions prevent farmers from applying a different adaptation plan, such as ecosystem-focused adaptation and implementing drought-tolerance crops. Information restrictions (for example, inadequate climate science information), governance restrictions (including communication disengages between national, local and community levels) and human capacity limitations (restricted capacities to examine dangers and effects) are acknowledged as dangerously influencing the application of adaptation plans. Cultural/social limitations (cultural status, class and gender) equally influence adaptation in frameworks with inherent traditions.
Asia	Human capacity, information, governance, and financial limitations normally contribute to the hindrances to urban adaptation. Financial, governance, economic and informational restrictions are associated with both soft and hard limits to adaptation against a series of risks in South Asia, whereas in West Asia, physical limitations to drought and heatwaves have been related to restrictions to adaptation.
Australasia	Several limitations, such as awareness, information, governance, cultural/social, human capacity, and economics, have been acknowledged as hindering adaptation action in the region. Indications of restrictions to adaptation are mostly for ecosystems, while communities and individuals are equally nearing soft restrictions due to social constraints.
Central and South America	Economic, governance, information, biophysical and cultural/social limitations are acknowledged as the greatest substantial factors in adaptation. Soft restrictions are mostly connected to governance limitations, although an indication of hard restrictions is associated with biophysical restrictions, such as glacier reduction causing loss of incomes and cultural values.
Europe	Major limitations are recognised as social, biophysical, technical and financial. For urban residents and main infrastructure, technical, environmental, and socioeconomic control limitations may cause restrictions on various spatial scales. Biophysical limitations may bring about restrictions to the ability of water saving and water efficiency measures to avoid water uncertainty under high warming states.
North America	Cultural/Social, governance, economic, information and biophysical limitations are recognised as the greatest significance for adaptation and causing soft and hard restrictions.
Small islands	Economic, governance, knowledge/awareness, cultural, technological and human capacity limitations are recognised as influencing adaptation and causing soft restrictions. Differences between limitations and soft limits in the small island framework is marginal, with policymakers in the Indian and Caribbean Oceans perceiving these as identical.

smallholder farming homes, Indigenous societies dependent on agriculture, and susceptible demographic regions, including children, women, the elderly, and socially secluded people. Coastal societies, particularly those reliant on fisheries, face noticeable income dangers, increasing inequalities within low-income populations. Moreover, health-associated risks are inclined to be more severe within low-income groups with inadequate infrastructure and housing access.

TABLE 7.2
Signs of Climate Change Implications Influencing the Availability of Economic Resources

Africa	Adverse effects on financial development and GDP growth rate are due to high usual temperatures and lesser rainfall. Financial damages from loss to infrastructure in the transport, energy, housing, water supply, communication services, education, and health sectors (observed)
Asia	High coastal losses are caused by the increase in sea level (Japan, India, China, Russia, and Korea). Decrease in aquaculture production. Damage to coastal ecosystem facilities (Bangladesh).
Australasia	Damage of wealth and adverse effects on GDP. High catastrophe prices (noticed in New Zealand and Australia).
Central and South America	High prices of severe measures relative to GDP (noticed in Belize, Guatemala). The decline in the growth of overall GDP per capita and overall earnings and labour earnings from one standard deviation in the strength of a hurricane windstorm
Europe	The adverse joint impact of numerous threats on the economy for Europe overall. The joint adverse effects of multiple threats on the economy of Southern Europe. High financial prices in agriculture and construction following flooding and heatwaves.
North America	A slight but persistent negative economy-wide effect exists on GDP (observed in the USA and Mexico). Economic risks associated with high-temperature scenarios. Trivial but determined encouraging economy extensive impact on GDP (noticed in Canada). Important financial prices for natural, urban and ecosystem facilities (USA). High financial losses for a subgroup of areas from high warming (southern and southeastern USA). Expensive responsibilities and the interruption of economic markets cause negative impacts on municipal budgets.
Small islands	High financial charges in relation to GDP from severe measures, mostly tropical cyclones. Adverse long-standing effects of severe actions for national budgets. Inundation of nearly every harbour and port service (Caribbean).

7.3 FACTORS DRIVING THE NATURAL RESOURCES TO EXPOSURE AND VULNERABILITY

The effects, threats, and responses to climate change arise in a context formed by many factors driving exposure and vulnerability. These consist of demographic changes, socio-economic growth (which includes inequalities), and the degradation of ecosystems. Moreover, larger worldwide trajectories, including technological developments, changes in global power dynamics, and resource deficiency, influence climate-connected hazards (Retief et al., 2016). A wider climate plan that deliberates power connections and social structures interrelating with climate change excesses is essential to comprehend the uneven effects of climate hazards, detected future losses, and vulnerability, as depicted in Figure 7.2.

Remarkably, these trajectories can both increase and moderate exposure and vulnerability. For instance, growing earnings, savings, and instituting social safety systems can improve robustness in the face of environmental shocks and pressures. The knowledge behind the drivers and future trajectories of vulnerability and exposure,

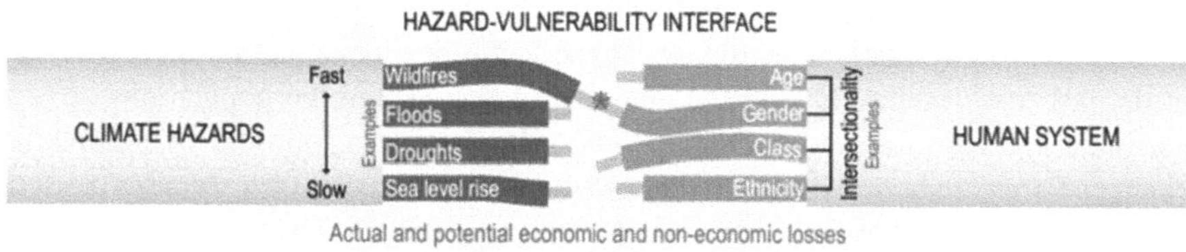

FIGURE 7.2 The relationships between climate hazards and human systems resulting in universal vulnerability.

combined with climate-induced variations in natural threats, is critical for complete risk evaluations and the advancement of efficient adaptation approaches. However, the empirical study proposes that these issues are regularly ignored in national adaptation strategic efforts (Garschagen et al., 2021). Despite being normally deliberated in separation, these drivers are complicatedly interrelated, functioning around both human and ecological systems and gradually stimulated by climate-connected threats and responses (Simpson et al., 2021). Besides, climate effects undermine these drivers and may compete with economic resources that could be assigned to development, moderation, adaptation, and resilience-building efforts, worsening the existing inequalities (Taconet, Méjean and Guivarch, 2020).

7.3.1 Demographics

Demographics are essential in determining the risks to natural resources and the dynamics of human settlements. An increase or decrease in population directly influences the availability of natural resources, including food, soil, energy, water, and fish stocks (Bongaarts, 2019). This trajectory may also give rise to the expansion or contraction of thickly inhabited regions (Day et al., 2016), with urban areas expected to occupy most of the world's population in the forthcoming decades. Urbanisation, while contributing to several advantages such as improved education and healthcare, upgraded infrastructure, and job opportunities, can create challenges such as resource shortage and increased pollution levels if not appropriately controlled (Rode et al., 2015). This trend increases exposure to numerous dangers. Remarkably, almost 10% of the world's population resided in low-lying coastal regions (less than 10 metres in elevation) in 2000 (Neumann et al., 2015), which is predicted to rise to 13.6% by 2100, depending on population situations (Neumann et al., 2015). The building of assets and facilities in naturally hazard-prone regions is equally expected to rise (Magnan et al., 2019), perhaps intensifying environmental degradation, such as the damage of flood-buffering wetlands (Oppenheimer et al., 2019). Moreover, demographic changes, such as fluctuations in income levels and ageing populations, can affect land utilisation patterns and the need for resources such as food, energy, and water (Arneth et al., 2019). Simultaneously, rural areas facing population decrease may experience sharp vulnerability and exposure, while ageing populations may increase susceptibility to certain climate-related risks, particularly heat stress (Byers et al., 2018).

7.3.2 Biodiversity and Ecosystems

The rapid escalation of human impacts on world biodiversity and ecosystems, mostly in the last five decades, has resulted in a sharp decrease in the populations of several wild species in the ocean and on land. This has been followed by substantial changes to terrestrial landscapes for agricultural purposes and the extensive production of strange and aggressive species (Bongaarts, 2019). Consequently, there is an increasing threat concerning the ability of ecosystems to maintain human society. For example, the percentage of all primary production used for human demands doubled during the 20th century (around 25% in 2005), although at a sluggish rate than the rise in human population (Krausmann et al., 2013). Forthcoming predictions centre meaningfully on bioenergy production, highlighting one of the interrelationships between responses to climate change and related risks (Apeh and Nwulu, 2024).

7.3.3 Trends in Poverty and Socio-economic Inequalities within and across Societies

Poverty is essential in increasing exposure and vulnerability by restricting households, individuals, and communities from accessing economic resources and hindering their adaptive capacities, including securing food and energy supplies or financing adaptive approaches (Hallegatte and Rozenberg, 2017). Until the beginning of the COVID-19 pandemic, there was an outstanding decrease in world poverty levels over the last decades. From 1981 to 2015, the proportion of the world population living in abject poverty (under the global poverty threshold of USD 1.90 per day) dropped from 42% to 10%, leaving around 736 million people in severe poverty, primarily concentrated in sub-Saharan Africa and South Asia (Bank, 2016). This universal decrease in poverty globally coincides with a reduction in vulnerability to several climate change influences (medium confidence).

Nevertheless, the COVID-19 pandemic triggered a substantial rise in abject poverty, affecting roughly 100 million persons in 2020, with uneven financial consequences on the most vulnerable, poorest, and least developed countries (Bank, 2020), thus increasing vulnerability to climate change. Several impoverished people are pastoralists and smallholder farmers whose incomes critically rely on climate-delicate natural ecosystems, mainly through semi-subsistence agriculture, where domestic food use depends largely on their food production (Mbow et al., 2019). A significant portion of this demographic experiences the extra threat of armed conflict, impeding financial development and growth while increasing local dependence on subsistence agriculture (Braithwaite, Dasandi and Hudson, 2016; Gieseken and Murphy, 2023), thereby exacerbating humanitarian crises and infuriating philanthropic challenges (Gieseken and Murphy, 2023).

Severe weather actions, mainly droughts, can push people into poverty cycles or force them more into hardship, increasing inequalities within and among countries. Climate threats are closely connected with other inequalities, regularly but not constantly interconnecting with poverty. The Fifth Assessment Report (AR5) of the United Nations Intergovernmental Panel on Climate Change (IPCC) concluded with high confidence that inequalities in vulnerability and exposure originate from complicated inequalities, commonly arising from irregular developing processes (Panel, 2013). These inequalities include political, geographic location, economic and socio-cultural factors, for example, religion, wealth, race/ethnicity, education, age, disability, gender, class/caste, and health status (Oppenheimer et al., 2019).

Since the AR5 research, several reports have restabilised and advanced this analysis, mostly concerning poverty and socio-economic inequality (Hallegatte and Rozenberg, 2017; Pelling and Garschagen, 2019). Economically underprivileged individuals are probably in exposed regions, including riverbanks or wastelands (Magnan et al., 2019). Moreover, poor individuals are excessively affected by climatic threats, obtain less assistance from their normal and equally penurious social networks following shocks, and often lack coverage from social safety programmes (Hallegatte and Rozenberg, 2017). Countries with high levels of inequality tend to show above-average exposure and vulnerability to climate-connected risks (Asadullah, Islam and Wahhaj, 2021).

Several socio-economic models utilised in climate studies have been criticised for their inadequate capacity to apprehend and represent poverty on a broader scale (Rao et al., 2019). Yet, a study of 92 countries discovered that comparative earning losses and other climate change effects excessively influence the poorest (Hallegatte and Rozenberg, 2017). Improvement has likewise been implemented in recognising and assigning the influences of climate change and vulnerability at the domestic level, mostly relating to women's activity and adaptive capacity (Rao et al., 2019). The distribution of effects and responses (both adaptation and moderation) affects inequality between countries and within them (Tol, 2021) and among various sections of societies. While distribution has mainly been assumed geographically, indicating those most in danger demand an extra emphasis on the social distribution of effects, responses, and resilience, for example, by different social safety coverage (Tenzing, 2020). Several climate responses interrelate with these universal threat drivers, some of which increase extra fairness concerns by sidelining the most vulnerable and worsening social conflicts (Oppenheimer et al., 2019), prompting broader investigations into the authority of climate threats (and impacts) across different scales. Therefore, the analysis of influences, responses, and threats is accompanied by evaluations of governance and the favourable environment for risk control.

7.4 OPPORTUNITIES FACILITATING CLIMATE-RESILIENT DEVELOPMENT FOR ADAPTATION AND SUSTAINABLE DEVELOPMENT

The correlation between climate change and development received little broad assessment until the first decades of the 21st century. Recently, mitigation and adaptation, the basic strategies for addressing climate change, have been treated as separate units in policy discussions and scientific research (Landauer, Juhola and Söderholm, 2015). However, current insights suggest that an integrated method, considering both adaptation and mitigation concurrently, could produce synergistic profits and reduce trade-offs (Runhaar et al., 2018). While some global involvements have proved synergies between adaptation and mitigation, challenging evidence-based assessments

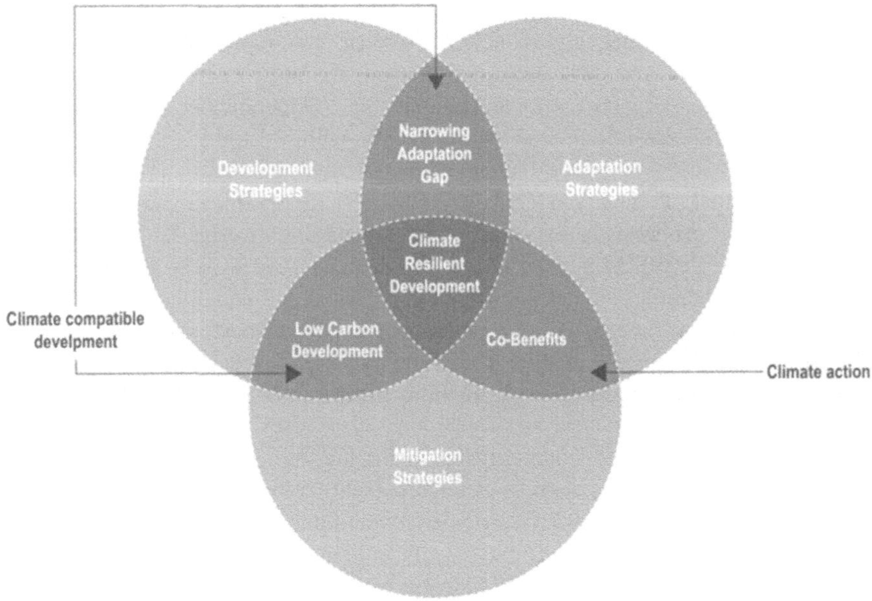

FIGURE 7.3 Climate resilient development (CRD).

of these synergies and trade-offs remain scarce. A crucial change in attention has been towards assessing climate-compatible development (CCD), described as development services that moderate climate change effects while boosting human development prospects in a low-emission and resilient future (Mitchell and Maxwell, 2010). CCD proposes triple wins, encouraging synergies between adaptation, mitigation, and development within singular intercessions, as shown in Figure 7.3. Mostly in low-income regions, CCD suggests strategic avenues for improving synergies between adaptation and mitigation.

Efficiently incorporating emission decreases and adaptive actions produces equally valuable results and holds the potential for cost efficiency and possibilities for sustainable development (Runhaar et al., 2018).

An evaluation of CCD strategies stresses the significance of purposely incorporating mitigation and adaptation procedures to accomplish climate security, meet fundamental demands, eradicate poverty, and promote equitable, fair, and sustainable development.

The literature on the relationship between adaptation, mitigation, equity, poverty, and sustainable development has advanced meaningfully (Pachauri et al., 2014). The IPCC Special Report on the effects of global warming of 1.5°C (Magnan et al., 2019) recommends that restricting warming to 1.5°C could synergistically align with poverty reduction, enhanced energy safety, and substantial public health advantages. Adopting the incorporated idea of CCD is estimated to yield transformative profits, mainly for the most vulnerable populations (Roy et al., 2018). However, threats related to transformative actions for the poor can be alleviated through effective governance across several levels and leveraging prevailing top-down and bottom-up processes, along with available policy levers, technology, education, and financial systems (Stringer et al., 2019).

7.4.1 Decision-making Approaches for Climate-resilient Development

A series of different traditional financial decision support devices is available to assist in assigning resources for climate change adaptation. Such tools include multi-criteria, cost-benefit, and cost-effectiveness analysis (Watkiss and Hunt, 2016). These devices primarily emphasise economic advantages and short-term perceptions. Moreover, there are procedures to estimate uncertainty, including iterative risk management, and steer up decision-making under uncertainty over longer time frames, such as real decision assessment and strong decision-making. Nevertheless, utilising these approaches needs expertise and human capital, typically concentrating on public instead of private adaptation decision-making processes. Economic-based methods regularly control the distributional effects among societal groups and fail to understand non-monetary aspects such as welfare, which are more abstract and challenging to measure (Stringer et al., 2019).

7.4.1.1 Policy Coherence, Policy Integration, and Broader Governance Approaches

Mainstreaming and coherence in policy stand out as critical approaches for handling both mitigation and adaptation in an integrated method, mainly based on effective governance. The dynamics of power, politics, and interests essentially affect the viability of incorporating climate policy with wider development purposes (Naess et al., 2015). The existence of institutional variations commonly results in inefficiencies and unproductive results (Di Gregorio et al., 2017). To facilitate higher comprehensive institutions and synergies, four main enabling situations have been acknowledged: (a) prearranged and current state laws, rules, and policies, (b) current and arranged economic resources and measures, (c) institutional preparations in the country with exact orientation to climate change challenge, and (d) prearranged and current programmes and enterprises in the country (Kabisch et al., 2016).

The alternative proposed approach comprises implementing a "dual track method" at the local level by adopting both local climate strategies and mainstreaming approaches (Duguma et al., 2014). This method can improve the adoption of climate activities and enable the distribution of climate challenges into local policymaking (Reckien et al., 2019). Efficient climate strategy incorporation demands consistency across four levels: internal (within moderation and adaptation strategies), external (between climate change and development purposes), vertical (integration into sectoral plans), and horizontal (establishing overarching governance structures) (Di Gregorio et al., 2017).

Development towards strategy integration differs across various levels, with mainstreaming and coherence advancing universally and steadily penetrating national contexts (Di Gregorio et al., 2017). Efforts to mainstream adaptation and moderation into food security agendas are crucial, demanding extensive oversight to incorporate food security, land restoration, disaster risk management, and climate rule into an integrated policy agenda (Woolf et al., 2015). However, adaptation and mitigation efforts in segments such as energy and agriculture present promising opportunities for incorporation into national plans (Antwi-Agyei et al., 2018). This emphasises the potential for comprehensive and sustainable financial and social expansion through coherent, cross-sectoral methods that aid in adaptation, mitigation, and development.

Several governance methods, including multi-level governance, adaptive governance, and polycentric governance, are gradually being used to pilot the change towards climate-resilient development (CRD). Polycentric governance, in particular, authorises local, regional, national, and international performers to jointly follow climate objectives through communication, trust-building, and strategy experimentation (Keating et al., 2014). This method highlights the importance of dispersed power and learning across numerous levels and performers. It is the source of power to accomplish joint objectives. Polycentric performers have the enclosing power, practical power, and power by design (Morrison et al., 2017). Polycentric governance contributes new prospects for climate activities through higher prospects for learning, communication, policy experimentation, and trust-building (Keating et al., 2014). Adaptive governance comprises connections among various performers and institutions to direct social-ecological systems towards anticipated states, which depend on nested institutions and multiple systems to adopt resilience and adaptation (Chaffin, Gosnell and Cosens, 2014). Community-based enterprises indicate the growth of adaptive governance in handling climate change challenges.

7.4.2 Community-Based Method

Applying a community-based method is essential to evaluating the implications of adaptation and mitigation actions while concurrently developing inclusivity, transparency, and accountability. This approach directly addresses adaptation and mitigation efforts and bolsters community capacity and social capital. For instance, in Bangkalan, Indonesia, strong social capacity and access to free agricultural resources are influential factors in efficient climate change adaptation and mitigation, alongside promoting community living (Sunkar and Santosa, 2018).

There is a rising demand to support the integration of Indigenous knowledge and the participation of Indigenous communities in strengthening unified policies in adaptation and mitigation (Ford et al., 2016). An in-depth understanding of local socio-ecological frameworks can produce the possibility to accelerate transformative methods that influence synergies (Thornton and Comberti, 2017). Research conducted in Ukraine focused on cooperatives highlights its effectiveness as a well-established livelihood policy encouraging small-scale agriculture. Besides, social capital is essential in capacity building and applying sustainable actions.

The Proklim plans exist in Indonesia as a distinguished enterprise concentrated on community-led adaptation and mitigation efforts. This enterprise authorises communities to understand climate change effects, collect data, and devise action plans (Muttaqin et al., 2019). Multi-stakeholder, participatory planning methods are influential in assisting farmers to evaluate and prioritise rural income policies in Indonesia. CRD is important across conservative development involvements surrounding regions such as water management, agricultural and aquacultural intensification and diversification, education, healthcare, food security, and farmer skill development (Wise et al., 2016).

7.5 ENABLING ENVIRONMENTS FOR ADAPTATION IN DIFFERENT SOCIO-ECONOMIC CONTEXTS

Climate change influences people differently, and everyone contributes disparately to climate change. A series of both economic and non-economic effects could be acknowledged. Thus, researchers support a rights-based adaptation strategy, mainly in addressing the marginalised and vulnerable (Bee, Biermann and Tschakert, 2013). United Nations treaty agencies have advised countries to maintain human rights obligations regarding climate change. This method, deep-rooted in human rights standards, emphasises transparency, non-discrimination, accountability, and participation across all levels of governance (Arts, 2017). Failure to tackle challenging interests during adaptation planning may result in futile results. This is observed in global efforts such as the Paris Agreement, where the principle of "common but separated duties and corresponding capacities" is dominant. A national-scale analysis reveals that low-income countries normally lack the needed economic, institutional, and financial capacity to manage climate change effects (Barr, Fankhauser and Hamilton, 2010). Issues including sanitation, insufficient water resources, governance failures, corruption, and lack of transparency worsen challenges in adaptation. The fundamental components that allow for adaptation to climate change in the environment are presented in Figure 7.4.

In contrast, higher-income countries profit from stronger institutions and better infrastructure, improving adaptation efficiency (Fankhauser and McDermott, 2014). Nevertheless, despite a single socio-economic background, several groups show diverse vulnerabilities and need personalised adaptation backing (Huynh and Stringer, 2018). For instance, Vietnam's households are vulnerable to sea-level increases, lack access to vital resources, and experience income uncertainties.

Public strategy aiming at climate change tries to address market failures, amend strategy misrepresentations, induce private adaptation, and offer climate services, climate-resilient safety nets, and public goods for the poor and vulnerable (Fankhauser, 2017). Producing and adopting such strategies is comparatively straightforward in countries with better institutional outlines. At the same time, those with weaker institutions, particularly those developing from conflict, may need more significant participation of regional financial commissions and international relationships to address "borderless climate risks" (Benzie and Persson, 2019), especially in places most prone to climate change effects. Supporting enabling environments in extremely susceptible and delicate states requires adapting funding and planning projects to encourage regional collaboration and transboundary adaptation. Even though such involvements may unintentionally strengthen main plans, dominating community voices and authority (Burch et al., 2019). Identifying the position of community empowerment is critical for effective vulnerability decrease and climate change adaptation.

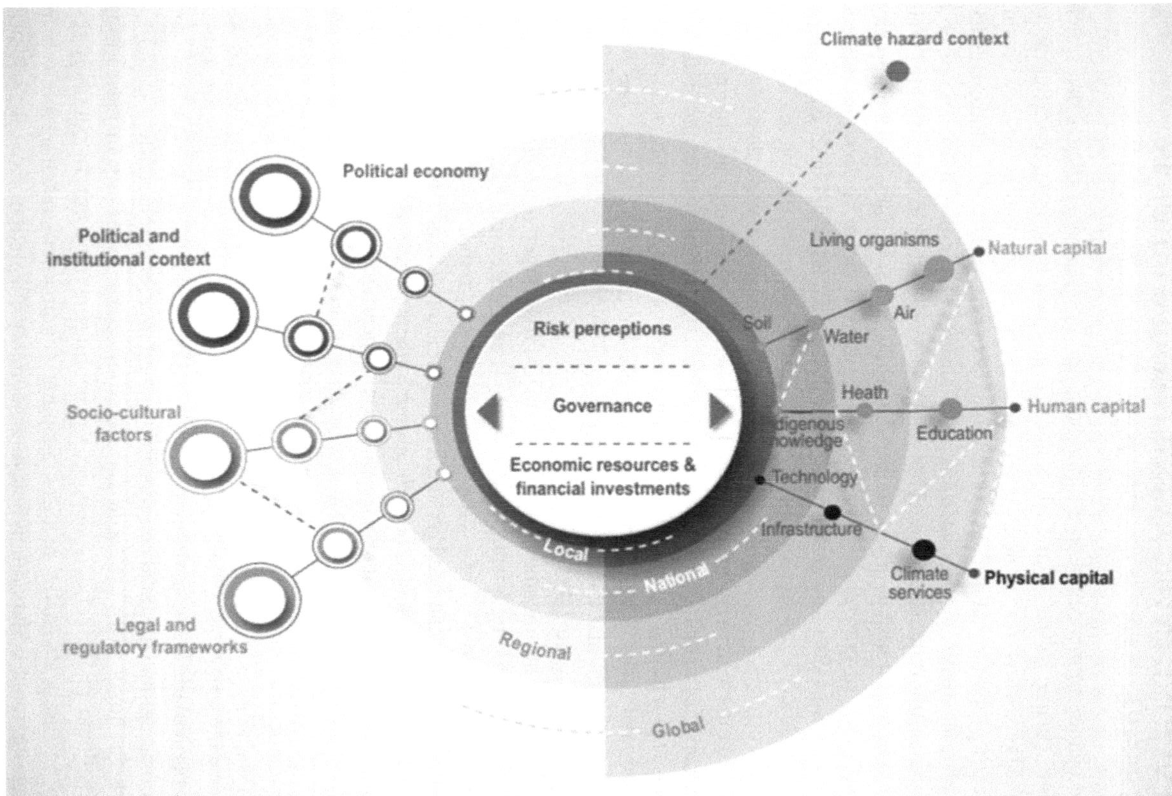

FIGURE 7.4 The fundamental components that allow for adaptation to climate change in the environment (Birkmann et al., 2022).

Moreover, regarding countries as separate units is insufficient because of complex international trade connections. For example, Europe has robust bonds with key trade associates such as Nigeria, India, Vietnam, and Indonesia, indicating that disregarding adaptation efforts elsewhere could expose vulnerabilities through unified supply chains (Lung, Füssel and Eichler, 2017). Strategies only protecting national interests, including food safety, may have contrary global effects, worsening prevailing inequities and excessively influencing countries and individuals least armed to manage climate change effects (Challinor, Adger and Benton, 2017). Behind every strategy lie decisions regarding the distribution of economic resources to exact adaptation measures, segments, and societal groups, both temporally and spatially. The IPCC evaluates that restraining global temperature increase to 1.5°C requires considerable yearly reserves in supply-side energy systems, from USD 1.6 trillion to USD 3.8 trillion between 2016 and 2050 (Mbow et al., 2019). Viewpoints on climate change threats influence resource distributions, the urgency of measures, motivational issues such as descriptive standards and supposed self-efficacy (Van Valkengoed and Steg, 2019), and divergent methods to appreciate human well-being, as demonstrated by Bhutan's Gross National Happiness framework (Kamei et al., 2021).

Nevertheless, growing economic mobilisation does not normally mean an efficient adaptation intervention on the ground, nor does it assure its efficiency. Unintentional adverse effects may occur because of a lack of knowledge of vulnerability drivers (e.g., gender inequality or unbalanced resource access), inadequate participation of marginalised local groups, incorporating adaptation into prevailing development programmes, and ineffectively describing adaptation accomplishment (Eriksen et al., 2021). A study in 2017 established that below 10% of climate investment assigned from international, regional, and national climate reserves to developing countries between 2003 and 2016 went to locally concentrated plans, signifying a demand for reassessment to confirm that the most affected groups can shape adequate resilience to climate change effects (Soanes et al., 2017).

7.5.1 Natural Capital

The intensifying impacts of climate change on environmental stresses, degenerated by human activities, are well-documented. This ecological degradation poses heavy challenges to adaptation efforts, mainly influencing vulnerable societies dependent on natural resources. Efficient management of natural capital is vital for strengthening resilience and improving ecosystems' capacity to acclimate to climate change, as emphasised by different studies (Pachauri et al., 2014; Mbow et al., 2019). Some systems, including mangroves in numerous countries, particularly developing countries, are essential in providing various ecosystem services. These services vary from coastal protection against sea-level rise and storm flows to carbon sequestration while boosting incomes through provisioning services, including timber and fisheries (Hamza et al., 2020). With the growing incidence of severe weather events because of climate change, conserving natural resources, including mangroves, has become important for coastal defence and adaptation. While not reducing the danger, the mangroves lower exposure and, in some circumstances, also reduces vulnerability.

The literature suggests that environmental assets reduce climate change effects and enable adaptation, mostly among marginalised communities deeply dependent on natural resources (Angelsen et al., 2014). However, the existing market economy does not typically explain the accurate value of natural assets, resulting in degradation and worsening climate change impacts. At the same time, growing populations, expanding consumption choices, and capitalist systems are putting pressure on natural resources. This challenge is compounded by short-term approaches that hinder efforts to alleviate poverty, even though addressing poverty is essential for human survival. Hence, all these features interrelate with the collective impact of deteriorating the effects of climate change and discouraging prospective adaptation choices, mainly for people in need (Apeh and Nwulu, 2025). Adaptation strategies occasionally deteriorate environmental degradation, such as constructing coral sea walls in Melanesia or developing agriculture in forested regions in the Congo Basin (Martin and Watson, 2016). This trajectory risks biodiversity, livelihoods, and food security, mainly among resource-reliant populations. Similarly, agricultural land is degraded worldwide, ending adaptation and livelihood possibilities for the underprivileged, natural resource-reliant populations while endangering food security, biodiversity, and human health at broader ranges.

Creating an enabling environment for adaptation requires reserves to conserve natural resources, adopt degradation expenses, and develop solid legal and controlling agendas (Scholes et al., 2018). In the literature, nature-based solutions (NBS) and ecosystem-based adaptations are promising for climate resilience, with possible economic advantages (Seddon et al., 2020). The Global Commission on Adaptation supports USD 1.8 trillion between 2020 and 2030 for adaptation procedures, which can produce considerable revenues. NBS influence natural procedures combined with engineered solutions to moderate climate hazards and increase societal well-being. Examples include green infrastructure such as built wetlands and urban forests, which not only decrease flood hazards but also offer entertainment places and develop air quality (Emmanuel and Loconsole, 2015). However, extra research is required to comprehend the complete range of assistance and unplanned effects related to NBS and efficiently incorporate them into policy agendas (Frantzeskaki et al., 2019).

7.5.2 Human Capital

Efficient adaptation demands encouraging human capital and socio-economic competencies, including education, experience, migration, knowledge, well-being, and health. This provision allows individuals to contribute to development initiatives expressively. Investing in human capital and initiatives intended to improve human capacities to tackle climate change additionally strengthens adaptation efforts, contributing positively to a decrease in total disaster risk. Studies examining educational achievement trends dating back to 1970 highlight that advancing education levels among working-age individuals constantly determine global financial growth, emphasising the essential role of education. Education fosters sustainable development by encouraging empowerment and means of information, such as climate change, and is associated with other areas of human capital, including mortality and health (Samir and Lutz, 2017). Evidence supports the idea that education decreases vulnerability and improves adaptive capacity (Sharma, Patwardhan and Patt, 2013). Additionally, education impacts risk insight, knowledge distribution, and problem-solving skills. Local learning and native knowledge contribute meaningfully to adaptation efforts, although they are usually ineffectively incorporated into formal educational systems and may inadequately address developing climate-associated risks.

Similarly, education is crucial in determining livelihood opportunities, food security, and income levels. Climate change poses dangers to education and human capital, with higher temperatures adversely affecting educational outcomes. Research reveals that severe weather actions such as snowstorms interrupt learning, yielding long-lasting and multidimensional learning environments (Graff Zivin, Hsiang and Neidell, 2018). Besides research evaluating formal education, a huge body of studies has emphasised social learning and its purpose in creating adaptive ability through combined information production and reflexivity. Foreground processing is necessary for constant fluctuations in response to developing situations. This research identifies the prospect of joint information for co-constructing strategies and practising responses to intricate, multi-stakeholder environmental issues. It addresses both the need for and challenge of such non-leading values and knowledge and proficiency in adaptation policymaking, considering the work of power dynamics within (Fisher and Dodman, 2019). Also, an emerging organisation of indication connects to organisational learning and adaptation. Organisations' adaptive products, such as those of homes and personalities, do not function in a vacuum, with organisations' conducts formed by strategy and market circumstances, among other issues. Mudombi et al. (2017) emphasise that many challenges in their research in South Africa arise from insufficient governance, limited resources, knowledge/skill gaps, and political interruptions within groups, as well as stringent project timelines.

Adaptations that sustain well-being and human health need funds for physical resources and infrastructure connected to water and hygiene, mostly in fast urbanising places in the Global South, combined with exact pro-poor asset policies that address uneven impacts of climate change on women. Climate change leads to spread of vector-borne diseases, malaria, and infections, such as meningitis (Rocklöv and Dubrow, 2020). Health effects are also caused by food uncertainty, which is caused by climate change, such as undernourishment and loss of income, making affordability and healthcare access harder. The main focus of human capital is local information on climate hazards, which consists of information systems outside Western scientific methods. However, deep-rooted in culture and passed down through generations, indigenous knowledge systems produce vital understandings of disaster prevention, moderation, and adaptation. Hitherto, they normally experience challenges in identification and conservation (Ajayi and Mafongoya, 2017). Efforts to incorporate native information into adaptation policies are crucial for improving resilience at local levels. Prioritising the presence of native information in climate evaluations and strategy developments can improve adaptation preparation and policymaking. Collaboration between traditional and scientific knowledge systems is indispensable for growing efficient adaptation policies that resonate with local settings. Solidifying plan agendas to simplify the incorporation of local and scientific knowledge is important for progressing adaptive capacity (Pachauri et al., 2014).

7.5.3 Physical Capital

Ensuring adequate investment in physical capital is crucial for advancing development programmes at the national level. However, for the underprivileged and marginalised society, physical capital is vital for adapting to several challenges (Hallegatte, Rentschler and Rozenberg, 2019). Physical capital involves assets such as land, infrastructure such as roads, water, mobile connectivity, housing, electricity supplies, and materials necessary for livelihoods, including forestry, farming, and fishing equipment, in addition to transportation and technology. Besides, it develops a sense of belonging and support for security. The effects of climate change on physical capital are generally wide, accompanied by substantial financial and emotional costs, mainly when societies experience difficulty because of inadequate sustainable human development, insufficient access to essential services, and income opportunities (Patel, 2018). Given the considerable investments essential to create and sustain physical capital at the state level, it is important to incorporate climate resilience deliberations into policymaking procedures, as reconstruction and replacement are both excessively costly. According to the World Bank, acclimatising from 2010 to 2050 to a world that is expected to be 2°C warmer by 2050 could cost between USD 70 billion and USD 100 billion annually, with the facility sector bearing the effect of these expenses (Bank, 2020). However, preventive measures can produce huge savings, with every USD 1 invested possibly avoiding USD 5 in repair costs (Dornan et al., 2013).

Successful adaptation policies associated with physical capital need suitable funding and technical skills, advanced planning, climate risk analysis, and healthy legal and institutional backgrounds. This comprises developing and implementing building codes and regulations, establishing insurance possibilities, and planning regulations intended to mitigate construction in zones extremely exposed to climate risks. Regrettably, on several occasions, these crucial enablers are missing. For instance, low-lying Least Developed Countries (LDCs) like Bangladesh and Small Island Developing States (SIDS) frequently endure climate-associated disasters, including hurricanes, cyclones, floods, typhoons, and saline intrusion, addressing the urgent demand for complete adaptation measures. Typhoons and related risks impose significant harm, disordering mobility, connectivity, and vital services like communication, food, energy, and water supplies, frequently leaving people homeless and deprived of their livelihood assets. Without suitable legal and institutional provisions, sustaining physical capital becomes more challenging.

Technology progressively assists climate change adaptation as a form of physical capital, although its distribution among uncertainty initiates fresh dangers such as hacking. Furthermore, efficient technology utilisation is associated with other types of capital, mainly human capital, demanding appropriate institutional support and training. Access to finance is crucial, with certain technological adaptations dependent on current facility and literacy levels, raising inequality concerns (Taylor, 2018). Studies emphasise the differential effects of technological change on genders and social classes, addressing the significance of highlighting fundamental inequalities to stop vulnerability.

Thorough testing of technologies across many frameworks is important to avoid maladaptation at scale while equally ensuring cultural sensitivity. A study in the

European Arctic with the Indigenous Sami Peoples illustrates how GPS technology and supplementary feeding can benefit some herders. However, there are concerns that overreliance on such technologies may diminish traditional skills and cultural knowledge over time (Andersson and Keskitalo, 2017). Generally, technological and other adaptations should aim to preserve the adaptive capacity of indigenous cultures like the Sami (Risvoll and Hovelsrud, 2016), mostly considering that reindeer grazing contributes to climate change mitigation efforts by protecting tundra ecosystems and enhancing surface albedo (Jaakkola, Juntunen and Näkkäläjärvi, 2018).

7.6 PRIVATE SECTOR AS A MEASURE IN CLIMATE CHANGE ADAPTATION

From the private sector's perception, different motivating factors prompt businesses to contribute to climate change adaptation efforts. This participation can be grouped into three major types, as outlined in Figure 7.5.

1. Ensuring business resilience along their value chains by adapting to climate change, thus reducing risks.
2. Facilitating the development of climate resilience among other stakeholders, such as civil society, businesses, and the public sector, through establishing financial resources and innovating products and services, hence exploiting an opportunity.
3. Maintaining compliance with regulatory agendas, incentive programmes, and external motivators, including preserving company reputation and responding to customer and community potentials.

7.6.1 Personal Decision to Support Climate Change-related Risks

From the perception of private sector interests, the present indication highlights different convincing motives to assess investment in adaptation as a requirement and an advantageous opportunity. Allocating resources towards climate resilience can now protect companies from increasing expenses related to climate change (Singh and Goyal, 2023). Furthermore, disregarding integrated adaptation and resilience considerations in investments can make companies prone to substantial threats, leading

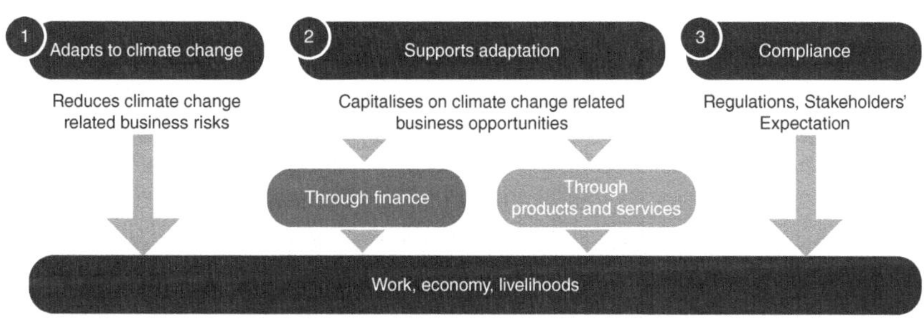

FIGURE 7.5 The motivation of private sector in climate change adaptation (Cochu, Hausotter and Henzler, 2019).

to unsuccessful investments, untimely physical damage, and deterioration, eventually affecting financial returns negatively. Moreover, there is a risk of losing vital business insurance coverage. Investors are gradually identifying this reality. The United Nations Environment Programme (UNEP) has conducted an analysis outlining several climate risks for the private sector, describing direct and indirect risks to businesses (Druce et al., 2016).

7.6.2 Identification of Business Opportunities Related to Climate Change

Developing and providing services and products pitched towards climate change adaptation can meaningfully impact several markets, resulting in growth, reductions, or new market opportunities. For example, in agriculture, there may be increased demand for irrigation systems, climate-resilient crops, farming technologies encouraging climate-smart activities, pest control products, and data services. In the same way, the health sector may require advanced technologies for quick diagnostics. Financial services can equally take advantage of these developments. Banks, insurance companies, and microfinance institutions can grab opportunities by producing financial possibilities for businesses to improve their resilience or capitalise on climate-related innovations. Besides, they can serve a critical function in raising awareness among private sector entities about the risks and opportunities related to climate change, hence encouraging requests for finance products. Adapting current public or private infrastructure can additionally initiate the demand for appropriate private sector services and skills for project application or co-financing. Such infrastructure enterprises may include climate-proofing production facilities, constructing flood-resistant roads, developing climate-resilient water treatment and distribution systems, or expanding healthcare facilities to accommodate potential patient rises quickly.

7.6.3 Reinforcement of Adaptation Finance

Governments globally are energetically pursuing measures to incorporate environmental considerations into their financial systems. This development is noticeable by the rising importance of unveiling climate-connected risks voluntarily or through mandatory agendas. Organised by the EU, this inclination is gaining ground globally. Several enterprises are already underway to substitute the development of a climate-conscious economic market. One prominent instance is the UN Task Force on Climate-Related Financial Disclosures (TCFD), which tries to regulate voluntary disclosures regarding climate risks. These discoveries are planned for implementation by companies, banks, and investors to improve stakeholder transparency.

Agendas such as the Social Bond Principles, Green Bond Principles, and Sustainability Bond Guidelines have developed leading standards for the global distribution of environmentally friendly, socially responsible, and sustainable bonds. Moreover, the Sustainable Banking Network, helped by the International Finance Corporation (IFC), assists in controlling agencies and banking associations from developing markets to improve their performance on social, environmental, and governance standards, such as climate risk organisations. Central banks and supervisors, mainly those associated with the Network for Greening the Financial System (NGFS), are crucial in considering climate-related risks within the financial sector. Therefore, financial institutions proactively develop climate risk management skills by engaging in innovative methodologies and tools. The application of labelled financial instruments, including green bonds and loans, as well as sustainability-linked bonds and loans, is facing substantial progress as avenues for financing climate-associated projects and activities.

7.7 CONCLUSION

This chapter discusses significant risks related to regions and natural resources due to climate change. It emphasises dangers across different sectors, regions, and scales, demonstrating vulnerability mostly in biodiversity hotspots. Vulnerability assessments from the INFORM Risk Index and WorldRiskIndex indicate high vulnerability in certain areas, particularly Africa, Asia, and Central America. Challenges, including displacement, inequality, and governance issues, limit adaptation capacities. Ecological disorders such as changes in food patterns, tree mortality, and global threats such as river flooding and heat waves have cascading impacts on incomes and food safety. Economic limitations hamper adaptation, especially in low- to middle-income regions. Climate change consequences, such as high coastal losses and reduced aquaculture production, disturb economic resources across regions. Certain parts, such as coastal regions and small islands, are hotspots for threats that affect human health, ecosystems, incomes, and infrastructure. Demographic changes impact natural resource availability and upsurge exposure to threats, with urbanisation worsening challenges like resource deficiency and pollution. Biodiversity loss threatens ecosystem sustainability, while poverty and socio-economic inequalities limit adaptive capacities, mainly in vulnerable populations. The chapter highlights the requirement for an integrated method for climate change and development, stressing climate-compatible development (CCD) as an approach to tackle adaptation and mitigation while encouraging sustainable development. Synergies between adaptation, mitigation, and growth can result in cost efficiency and sustainable outcomes, mostly in low-income regions.

REFERENCES

Ajayi, O. C. and Mafongoya, P. L. (2017) *Indigenous knowledge systems and climate change management in Africa*. CTA.

Andersson, E. and Keskitalo, E. C. H. (2017) 'Technology use in Swedish reindeer husbandry through a social lens'. *Polar Geography*, 40(1), pp. 19–34.

Angelsen, A. et al. (2014) 'Environmental income and rural livelihoods: A global-comparative analysis'. *World Development*, 64, pp. S12–S28.

Antwi-Agyei, P. et al. (2018) 'Alignments between nationally determined contributions and the sustainable development goals for West Africa's'. *Climate Policy*, 18(10), pp. 1296–1312.

Apeh, O. O., & Nwulu, N. (2024). The Food-Energy-Water Nexus Optimization: A Systematic Literature Review. *Research on World Agricultural Economy*, 5(4).

Apeh, O. O., & Nwulu, N. I. (2025). Machine learning predictions for fault detections in solar photovoltaic systems: A bibliographic outlook. *Journal of Infrastructure, Policy and Development*, 9(2), Article 9940.

Arneth, A. et al. (2019) 'Framing and context'. In Priyadarshi R. Shukla, James Skea, E. Calvo Buendia, Valérie Masson-Delmotte, H.-O. Pörtner, D.C. Roberts, P. Zhai, R. Slade, S. Connors, R. van Diemen, M. Ferrat, E. Haughey, S. Luz, M. Pathak, J. Petzold, J. Portugal Pereira, P. Vyas, E. Huntley, K. Kissick, M. Belkacemi & J. Malley (eds.) *Climate change and land: An IPCC special report on climate change, desertification, land degradation, sustainable land management, food security, and greenhouse gas fluxes in terrestrial ecosystems* (pp. 1–98). Intergovernmental Panel on Climate Change (IPCC).

Arts, K. (2017) 'Inclusive, sustainable development: A human rights perspective'. *Current Opinion in Environmental Sustainability*, 24, pp. 58–62.

Asadullah, M. N., Islam, K. M. M. and Wahhaj, Z. (2021) 'Child marriage, climate vulnerability and natural disasters in coastal Bangladesh'. *Journal of Biosocial Science*, 53(6), pp. 948–967.

Bank, W. (2016) *Employment in agriculture (% of total employment)*. World Bank.

Bank, W. (2020) *Global economic prospects, June 2020*. The World Bank.

Barr, R., Fankhauser, S. and Hamilton, K. (2010) 'Adaptation investments: A resource allocation framework'. *Mitigation and Adaptation Strategies for Global Change*, 15, pp. 843–858.

Bee, B., Biermann, M. and Tschakert, P. (2013) 'Gender, development, and rights-based approaches: Lessons for climate change adaptation and adaptive social protection'. In Margaret Alston and Kerri Whittenbury (eds) *Research, action and policy: Addressing the gendered impacts of climate change* (pp. 95–108). Springer.

Benzie, M. and Persson, Å. (2019) 'Governing borderless climate risks: Moving beyond the territorial framing of adaptation'. *International Environmental Agreements: Politics, Law and Economics*, 19, pp. 369–393.

Birkmann, J. et al. (2022) *Poverty, livelihoods and sustainable development*. Cambridge University Press.

Bongaarts, J. (2019) *IPBES, 2019. Summary for policymakers of the global assessment report on biodiversity and ecosystem services of the intergovernmental science-policy platform on biodiversity and ecosystem services*. Wiley Online Library.

Braithwaite, A., Dasandi, N. and Hudson, D. (2016) 'Does poverty cause conflict? Isolating the causal origins of the conflict trap'. *Conflict Management and Peace Science*, 33(1), pp. 45–66.

Burch, S. et al. (2019) 'New directions in earth system governance research'. *Earth System Governance*, 1, p. 100006.

Byers, E. et al. (2018) 'Global exposure and vulnerability to multi-sector development and climate change hotspots'. *Environmental Research Letters*, 13(5), p. 55012.

Cansino-Loeza, B. and Ponce-Ortega, J. M. (2021) 'Sustainable assessment of water-energy-food nexus at regional level through a multi-stakeholder optimization approach'. *Journal of Cleaner Production*, 290, p. 125194.

Chaffin, B. C., Gosnell, H. and Cosens, B. A. (2014) 'A decade of adaptive governance scholarship: Synthesis and future directions'. *Ecology and Society*, 19(3), 56

Challinor, A. J., Adger, W. N. and Benton, T. G. (2017) 'Climate risks across borders and scales'. *Nature Climate Change*, 7(9), pp. 621–623.

Cochu, A., Hausotter, T. and Henzler, M. (2019) 'The roles of the private sector in climate change adaptation – An introduction'. *Extraído el*, 26.

Day, J. W. et al. (2016) 'Approaches to defining deltaic sustainability in the 21st century'. *Estuarine, Coastal and Shelf Science*, 183, pp. 275–291.

Di Gregorio, M. et al. (2017) 'Climate policy integration in the land use sector: Mitigation, adaptation and sustainable development linkages'. *Environmental Science & Policy*, 67, pp. 35–43.
Dornan, M. et al. (2013) *Infrastructure maintenance in the Pacific: Challenging the build, neglect, rebuild paradigm.* Pacific Regional Infrastructure Advisory Center.
Druce, L. et al. (2016) *Demystifying adaptation finance for the private sector.* The United Nations Environment Programme (UNEP).
Duguma, L. A. et al. (2014) 'A systematic analysis of enabling conditions for synergy between climate change mitigation and adaptation measures in developing countries'. *Environmental Science & Policy*, 42, pp. 138–148.
Emmanuel, R. and Loconsole, A. (2015) 'Green infrastructure as an adaptation approach to tackling urban overheating in the Glasgow Clyde Valley Region, UK'. *Landscape and Urban Planning*, 138, pp. 71–86.
Eriksen, S. et al. (2021) 'Adaptation interventions and their effect on vulnerability in developing countries: Help, hindrance or irrelevance?' *World Development*, 141, p. 105383.
Fankhauser, S. (2017) 'Adaptation to climate change'. *Annual Review of Resource Economics*, 9, pp. 209–230.
Fankhauser, S. and McDermott, T. K. J. (2014) 'Understanding the adaptation deficit: Why are poor countries more vulnerable to climate events than rich countries?' *Global Environmental Change*, 27, pp. 9–18.
Fisher, S. and Dodman, D. (2019) 'Urban climate change adaptation as social learning: Exploring the process and politics'. *Environmental Policy and Governance*, 29(3), pp. 235–247.
Ford, J. D. et al. (2016) 'Including indigenous knowledge and experience in IPCC assessment reports'. *Nature Climate Change*, 6(4), pp. 349–353.
Frantzeskaki, N. et al. (2019) 'Nature-based solutions for urban climate change adaptation: Linking science, policy, and practice communities for evidence-based decision-making'. *Bioscience*, 69(6), pp. 455–466.
Gain, A. K., Rouillard, J. J. and Benson, D. (2013) *Can integrated water resources management increase adaptive capacity to climate change adaptation? A critical review.* Scientific Research Publishing.
Garschagen, M. et al. (2021) 'The consideration of future risk trends in national adaptation planning: Conceptual gaps and empirical lessons'. *Climate Risk Management*, 34, p. 100357.
Gieseken, H. O. and Murphy, V. (2023) 'The protection of the natural environment under international humanitarian law: The ICRC's 2020 guidelines'. *International Review of the Red Cross*, 105(924), pp. 1180–1207.
Graff Zivin, J., Hsiang, S. M. and Neidell, M. (2018) 'Temperature and human capital in the short and long run'. *Journal of the Association of Environmental and Resource Economists*, 5(1), pp. 77–105.
Hallegatte, S., Rentschler, J. and Rozenberg, J. (2019) *Lifelines: The resilient infrastructure opportunity.* World Bank Publications.
Hallegatte, S. and Rozenberg, J. (2017) 'Climate change through a poverty lens'. *Nature Climate Change*, 7(4), pp. 250–256.
Hamza, A. J. et al. (2020) 'Past and present utilization of mangrove resources in Eastern Africa and drivers of change, *Journal of Coastal Research*, 95(SI), pp. 39–44.
Hoff, H. (2011) *Understanding the nexus.* Stockholm Environment Institute (SEI).
Hua, E. et al. (2021) 'Water competition mechanism of food and energy industries in WEF nexus: A case study in China'. *Agricultural Water Management*, 254, p. 106941.
Huynh, L. T. M. and Stringer, L. C. (2018) 'Multi-scale assessment of social vulnerability to climate change: An empirical study in coastal Vietnam'. *Climate Risk Management*, 20, pp. 165–180.

Jaakkola, J. J. K., Juntunen, S. and Näkkäläjärvi, K. (2018) 'The holistic effects of climate change on the culture, well-being, and health of the Saami, the only indigenous people in the European Union'. *Current Environmental Health Reports*, 5, pp. 401–417.

Kabisch, N. et al. (2016) 'Nature-based solutions to climate change mitigation and adaptation in urban areas: Perspectives on indicators, knowledge gaps, barriers, and opportunities for action'. *Ecology and Society*, 21(2) 39.

Kamei, M. et al. (2021) 'Urbanization, carbon neutrality, and gross national happiness: Sustainable development pathways for Bhutan's. *Cities*, 111, p. 102972.

Keating, B. A. et al. (2014) 'Food wedges: Framing the global food demand and supply challenge towards 2050'. *Global Food Security*, 3(3–4), pp. 125–132.

Krausmann, F. et al. (2013) 'Global human appropriation of net primary production doubled in the 20th century'. *Proceedings of the National Academy of Sciences*, 110(25), pp. 10324–10329.

Landauer, M., Juhola, S. and Söderholm, M. (2015) 'Inter-relationships between adaptation and mitigation: A systematic literature review'. *Climatic Change*, 131, pp. 505–517.

Leck, H. et al. (2015) 'Tracing the water–energy–food nexus: Description, theory and practice'. *Geography Compass*, 9(8), pp. 445–460.

Liu, Y. and Chen, B. (2020) 'Water-energy scarcity nexus risk in the national trade system based on multiregional input-output and network environ analyses'. *Applied Energy*, 268, p. 114974.

Lung, T., Füssel, H. M., & Eichler, L. (2017). Europe's vulnerability to climate change impacts outside Europe. In *Climate Change, Impacts and Vulnerability in Europe 2016* (EEA Report 1/2017), Luxembourg: European Environment Agency. ISBN 978-92-9213-835-6.

Ma, Y. et al. (2021) 'Mathematical modeling for planning water-food-ecology-energy nexus system under uncertainty: A case study of the Aral Sea Basin'. *Journal of Cleaner Production*, 308, p. 127368.

Magnan, A. K. et al. (2019) *Low-lying islands and coasts*. Intergovernmental Panel on Climate Change (IPCC).

Martin, T. G. and Watson, J. E. M. (2016) 'Intact ecosystems provide the best defence against climate change'. *Nature Climate Change*, 6(2), pp. 122–124.

Mbow, C. et al. (2019) 'Food security'. In J. Skea, P. R. Shukla, E. Calvo Buendia, V. Masson-Delmotte, H.-O. Pörtner, D. C. Roberts, P. Zhai, R. Slade, S. Connors, R. van Diemen, M. Ferrat, Eamon Haughey, S. Luz, S. Neogi, M. Pathak, J. Petzold, J. Portugal Pereira, P. Vyas, E. Huntley, K. Kissick, M. Belkacemi, J. Malley (eds) *Climate change and land: An IPCC special report on climate change, desertification, land degradation, sustainable land management, food security, and greenhouse gas fluxes in terrestrial ecosystems*. Intergovernmental Panel on Climate Change.

Mitchell, T. and Maxwell, S. (2010) 'Defining climate compatible development'. *CDKN ODI policy brief* (pp. 1–6). CDKN London.

Mohtar, R. H. and Daher, B. (2016) 'Water-energy-food nexus framework for facilitating multi-stakeholder dialogue'. *Water International*, 41(5), pp. 655–661.

Molajou, A. et al. (2021) 'A new paradigm of water, food, and energy nexus'. *Environmental Science and Pollution Research*, 30, pp. 1–11.

Morrison, T. H. et al. (2017) 'Mitigation and adaptation in polycentric systems: Sources of power in the pursuit of collective goals'. *Wiley Interdisciplinary Reviews: Climate Change*, 8(5), p. e479.

Mudombi, S. et al. (2017) 'The use of and obstacles to social learning in climate change adaptation initiatives in South Africa'. *Jàmbá: Journal of Disaster Risk Studies*, 9(1), pp. 1–8.

Muttaqin, Z., Yulianti, A. and Karmanah (2019) 'Climate village program (ProKlim) in Simurugul Sub-Village, Margawati Village, Garut Kota Sub-Regency, Garut Regency, West Java Province, Indonesia'. In *IOP Conference Series: Earth and Environmental Science* (p. 12046). IOP Publishing.

Naess, L. O. et al. (2015) 'Climate policy meets national development contexts: Insights from Kenya and Mozambique'. *Global Environmental Change*, 35, pp. 534–544.
Neumann, B. et al. (2015) 'Future coastal population growth and exposure to sea-level rise and coastal flooding-a global assessment'. *PLoS One*, 10(3), p. e0118571.
Oppenheimer, M. et al. (2019) *The ocean and cryosphere in a changing climate*. Cambridge University Press.
Pachauri, R. K. et al. (2014) *Climate change 2014: Synthesis report. contribution of working groups I, II and III to the fifth assessment report of the intergovernmental panel on climate change*. IPCC.
Panel, I. (2013) 'The fifth assessment report (AR5) of the United Nations intergovernmental panel on climate change (IPCC)'. In T. F. Stocker, D. Qin, G.-K. Plattner, M. Tignor, S. K. Allen, J. Boschung, A. Nauels, Y. Xia, V. Bex, & P. M. Midgley (eds) *Climate change 2013: The physical science basis. Contribution of Working Group I to the Fifth Assessment Report of the Intergovernmental Panel on Climate Change. Climate change 2013: The physical science basis*. Cambridge University Press. https://doi.org/10.1017/CBO9781107415324
Patel, N. (2018) *Figure of the week: Understanding poverty in Africa*. Brookings Institution.
Pelling, M. and Garschagen, M. (2019) 'Put equity first in climate adaptation'. *Nature*, 569(7756), pp. 327–329.
Rao, N. D. et al. (2019) 'Income inequality projections for the shared socioeconomic pathways (SSPs)'. *Futures*, 105, pp. 27–39.
Reckien, D. et al. (2019) 'Dedicated versus mainstreaming approaches in local climate plans in Europe'. *Renewable and Sustainable Energy Reviews*, 112, pp. 948–959.
Retief, F. et al. (2016) 'Global megatrends and their implications for environmental assessment practice'. *Environmental Impact Assessment Review*, 61, pp. 52–60.
Risvoll, C. and Hovelsrud, G. K. (2016) 'Pasture access and adaptive capacity in reindeer herding districts in Nordland, Northern Norway'. *The Polar Journal*, 6(1), pp. 87–111.
Rocklöv, J. and Dubrow, R. (2020) 'Climate change: An enduring challenge for vector-borne disease prevention and control'. *Nature Immunology*, 21(5), pp. 479–483.
Rode, K. D. et al. (2015) 'Can polar bears use terrestrial foods to offset lost ice-based hunting opportunities?' *Frontiers in Ecology and the Environment*, 13(3), pp. 138–145.
Roy, J. et al. (2018) *Sustainable development, poverty eradication and reducing inequalities*. Cambridge University Press.
Runhaar, H. et al. (2018) 'Mainstreaming climate adaptation: Taking stock about "What Works" from empirical research worldwide'. *Regional Environmental Change*, 18, pp. 1201–1210.
Samir, K. C. and Lutz, W. (2017) 'The human core of the shared socioeconomic pathways: Population scenarios by age, sex and level of education for all countries to 2100'. *Global Environmental Change*, 42, pp. 181–192.
Scholes, R. et al. (2018) *Summary for policymakers of the assessment report on land degradation and restoration of the intergovernmental science-policy platform on biodiversity and ecosystem services*. IPBES Secretariat.
Seddon, N. et al. (2020) 'Understanding the value and limits of nature-based solutions to climate change and other global challenges'. *Philosophical Transactions of the Royal Society B*, 375(1794), p. 20190120.
Sharma, U., Patwardhan, A. and Patt, A. G. (2013) 'Education as a determinant of response to cyclone warnings: Evidence from coastal zones in India'. *Ecology and Society*, 18(2), 18.
Simpson, N. P. et al. (2021) 'A framework for complex climate change risk assessment'. *One Earth*, 4(4), pp. 489–501.
Singh, S. and Goyal, M. K. (2023) 'Enhancing climate resilience in businesses: The role of artificial intelligence'. *Journal of Cleaner Production*, 418, p. 138228.

Soanes, M. et al. (2017) *Delivering real change: Getting international climate finance to the local level*. International Institute for Environment and Development.

Stringer, C. et al. (2019) 'High-dimensional geometry of population responses in visual cortex'. *Nature*, 571(7765), pp. 361–365.

Sunkar, A. and Santosa, Y. (2018) 'Low resource use and high yield concept in climate-smart community empowerment'. In Benjamin McLellan (ed) *Sustainable future for human security: Society, cities and governance* (pp. 319–332). Springer.

Taconet, N., Méjean, A. and Guivarch, C. (2020) 'Influence of climate change impacts and mitigation costs on inequality between countries'. *Climatic Change*, 160(1), pp. 15–34.

Taylor, M. (2018) 'Climate-smart agriculture: What is it good for?' *The Journal of Peasant Studies*, 45(1), pp. 89–107.

Tenzing, J. D. (2020) 'Integrating social protection and climate change adaptation: A review'. *Wiley Interdisciplinary Reviews: Climate Change*, 11(2), p. e626.

Thornton, T. F. and Comberti, C. (2017) 'Synergies and trade-offs between adaptation, mitigation and development'. *Climatic Change*, 140, pp. 5–18.

Tol, R. S. J. (2021) 'The distributional impact of climate change'. *Annals of the New York Academy of Sciences*, 1504(1), pp. 63–75.

Van Valkengoed, A. M. and Steg, L. (2019) 'Meta-analyses of factors motivating climate change adaptation behaviour'. *Nature Climate Change*, 9(2), pp. 158–163.

Watkiss, P. and Hunt, A. (2016) 'Assessing climate-resilient development options'. In Sam Fankhauser and Thomas K.J. McDermott (eds) *The economics of climate-resilient development*. Edward Elgar Publishing.

Wise, R. M. et al. (2016) 'How climate compatible are livelihood adaptation strategies and development programs in rural Indonesia?' *Climate Risk Management*, 12, pp. 100–114.

Woolf, D., Jirka, S., Solomon, D., & Lehmann, J. (2015). *Climate change mitigation potential of Ethiopia's productive safety-net program (PSNP)*. Cornell University. Retrieved from https://ecommons.cornell.edu/items/39932423-2fae-4a82-ab5d-2849a04a662d?utm_source=chatgpt.com

Index

Pages in *italics* refer to figures and pages in **bold** refer to tables.

G

H

I

K

L

M

N

O

P

R

S

T

U

V

W

Z